Lecture Notes in Computer Science   1754
Edited by G. Goos, J. Hartmanis and J. van Leeuwen

Springer

Berlin
Heidelberg
New York
Barcelona
Hong Kong
London
Milan
Paris
Singapore
Tokyo

Jouko Väänänen (Ed.)

Generalized Quantifiers and Computation

9th European Summer School
in Logic, Language, and Information
ESSLLI'97 Workshop
Aix-en-Provence, France, August 11-22, 1997
Revised Lectures

 Springer

Series Editors

Gerhard Goos, Karlsruhe University, Germany
Juris Hartmanis, Cornell University, NY, USA
Jan van Leeuwen, Utrecht University, The Netherlands

Volume Editor

Jouko Väänänen
Department of Mathematics, University of Helsinki
P.O. Box 4, Yliopistonkatu 5, 00014 Helsinki, Finland
E-mail: jouko.vaananen@helsinki.fi

Cataloging-in-Publication data applied for

Die Deutsche Bibliothek - CIP-Einheitsaufnahme

Generalized quantifiers and computation : revised lectures / 9th European
Summer School in Logic, Language, and Information, ESSLLI '97, Aix-en-Provence,
France, August 11 - 22, 1997. Jouko Väänänen (ed.). - Berlin ;
Heidelberg ; New York ; Barcelona ; Hong Kong ; London ; Milan ;
Paris ; Singapore ; Tokyo : Springer, 1999
(Lecture notes in computer science ; Vol. 1754)
ISBN 3-540-66993-0

CR Subject Classification (1998): F.4.1, I.2.3

ISSN 0302-9743
ISBN 3-540-66993-0 Springer-Verlag Berlin Heidelberg New York

Typesetting: Camera-ready by author
SPIN: 10719588   06/3142 – 5 4 3 2 1 0   Printed on acid-free paper

Foreword

The papers contained in this volume were presented at the workshop *Generalized Quantifiers and Computation* of ESSLLI'97, the 9th European Summer School in Logic, Language and Information in Aix-en-Provence, France, August 11-22, 1997.

ESSLLI'97 was organised under the auspices of the European Association for Logic, Language and Information (FoLLI). The annual ESSLLI Summer School has developed into an important meeting place and forum for discussion for researchers and students interested in the interdisciplinary study of Logic, Computer Science, Linguistics, and Cognitive Science.

The purpose of the workshop *Generalized Quantifiers and Computation* in ESSLLI'97 was to bring together experienced researchers and students in the areas of generalized quantifiers in finite model theory on one hand, and regular languages and circuit complexity on the other hand.

The speakers at the workshop were:

1. Anuj Dawar: Finite variable equivalence and generalized quantifiers.
2. Kousha Etessami: Dynamic tree-isomorphism via first-order update.
3. Lauri Hella: Enhancing fixed point logic with generalized quantifier.
4. Christian Michaux: Definability and undefinability in extensions of Presburger arithmetic, Buchi arithmetics. Applications.
5. Ari Koponen: Definability of group theoretical notions.
6. Kerkko Luosto: Ramsey theory is needed for solving definability problems of generalized quantifiers.
7. Johann Makowsky: Invariant definability and circuit complexity.
8. Juha Nurmonen: Local properties and finite structures.
9. Iain Stewart: Logical characterization of oracle complexity classes.
10. Jouko Väänänen: Generalized quantifiers and computation - an introduction.
11. Helmut Veith: Generalized quantifiers in logic programming and query languages.
12. Heribert Vollmer: Generalized quantifiers in complexity theory.

I am grateful to all the authors who contributed a paper to this proceedings volume.

October 1999   Jouko Väänänen

Table of Contents

Generalized Quantifiers, an Introduction[*]

Jouko Väänänen

Department of Mathematics, P.O.Box 4 (Yliopistonkatu 5)
00014 University of Helsinki, Finland
jouko.vaananen@helsinki.fi
http://www.logic.math.helsinki.fi

Abstract. We review recent work in the field of generalized quantifiers on finite models. We give an idea of the methods that are available in this area. Main emphasis is on definability issues, such as whether there is a logic for the PTIME properties of unordered finite models.

1 Introduction

The ordinary quantifiers "for some" and "for all" are not sufficient for expressing some basic mathematical concepts. This led **Mostowski** [22] to introduce in 1957 *generalized quantifiers*, such as "for infinitely many" and "for uncountably many". In the 1960's and 1970's these and other similar quantifiers were intensively studied by logicians. In this decade generalized quantifiers re-emerged in the framework of finite structures. Researchers in *descriptive complexity theory* and *natural language semantics* were looking at ways to formalize expressions like "for at least half" and "for an even number". It turned out that the concept of generalized quantifier introduced by Mostowski, and further extended by **Lindström** [21], was the right answer.

The goal of descriptive complexity theory is to classify problems, not according to how much resources they need when solved by means of a Turing machine, but according to how powerful logical languages are necessary for describing the problems. For a quick example, let us consider some graph problems, say the problem whether a graph is triangle-free.

We are given a graph, that is, a pair (G, E), where G is a set and E is a binary predicate representing the edge-relation. Even before we think about triangle-freeness, we may ask how difficult it is to express the fact that (G, E) is indeed a graph. We have to check that E is symmetric and anti-reflexive. These two axioms can be easily expressed in a logical language as follows: In what follows, x and y are variables for elements of the domain G and $E(x, y)$ stands for the assertion that x and y are in the edge-relation. Using standard logical notation, $\forall$ means "for all", $\rightarrow$ means "implies", $\neg$ means "not" and $\wedge$ means "and", the following sentence expresses the fact that (G, E) is a graph:

$$\forall x \forall y (E(x, y) \rightarrow E(y, x)) \wedge \forall x \neg E(x, x).\psi \tag{1}$$

---

[*] Reprinted with permission from the Bulletin of the European Association for Theoretical Computer Science, 62 (1997), 115–136.

J. Väänänen (Ed.): ESSLLI'97, LNCS 1754, pp. 1–17, 1999.

Likewise, the sentence

$$\forall x \forall y \forall z \neg (E(x,y) \wedge E(y,z) \wedge E(z,x)). \tag{2}$$

expresses triangle-freeness of the graph (G, E). From the point of view of descriptive complexity theory we may now say that to express the graph-axioms, and also the triangle-freeness of a graph, we do not need more than a logical language which has variables for elements and permits the use of $\forall$, $\neg$, $\wedge$ and $\rightarrow$. The smallest such language is called *first order logic* FO.

The logical language FO has become an extremely successful tool in logic in this century. However, this success is almost solely due to its handiness in expressing completeness properties of infinite structures such as "for every number there is a bigger number" or "between any two distinct numbers there is another number" etc. Problems arising in computer science do not usually involve structures with universal completeness properties, not least because these structures (databases, computations, etc) are *finite*. Respectively, all the usual methods in the theoretical study of FO almost systematically fail in the finite context (see e.g. [11,12]). Indeed, first order logic is not at all handy in expressing interesting properties of finite structures. Let us take the path-problem as an example. From a programming point of view it is natural to express the existence of a path from x to y in a graph (G, E) by a formula $\mathrm{Conn}(x, y)$ such that

$$\mathrm{Conn}(x, y) \leftrightarrow [E(x, y) \vee \exists z (E(x, z) \wedge \mathrm{Conn}(z, y))]. \tag{3}$$

This sentence features $\leftrightarrow$ ("if and only if"), $\exists$ ("there exists") and $\vee$ ("or"), which are all definable in terms of the operations of FO. But more interestingly, it seems that in order to get the formula $\mathrm{Conn}(x, y)$ we have to "solve" the equivalence (3) since Conn occurs on both sides. Such solutions are called *fixed points*. In general there may be several different fixed points, but there is a simple way of guaranteeing that a unique least fixed point exists. In (3) the existence of a unique least fixed point characteristically follows from the fact that on the right hand side the predicate Conn occurs only positively. The smallest logic extending first order logic where such fixed points can be expressed is called *fixed point logic* FP. It was introduced by **Aho** and **Ullman** [1] in 1979. They also proved that the least fixed point of the equivalence (3) is not first order definable.

Fixed point logic is definitely very different from first order logic. Expressions of first order logic can be written down on a piece of paper, but how to write down the expression Conn? If we make a try, changing variables to avoid confusion, the result is likely to look like this: $\mathrm{Conn}(x, y) \leftrightarrow$

$$E(x, y) \vee$$
$$\exists z_1 (E(x, z_1) \wedge (E(z_1, y) \vee$$
$$\exists z_2 (E(z_1, z_2) \wedge (E(z_2, y) \vee \dots))).$$

This sentence is never-ending! The expressions of fixed point logic are not sentences in the ordinary sense of the word but some kind of self-referential recursive generalized sentences. Despite this difficulty in construing fixed point queries as sentences, fixed point logic has a very clear computational content. Whenever a graph and a fixed point expression is given, it is immediate how to check whether the graph satisfies the expression, and this can be done in polynomial time in the size of the graph.

It would be tempting to conjecture that not only is every fixed point query in polynomial time, but conversely every polynomial time graph property is expressible as a fixed point query. Indeed, **Immerman** [19] and **Vardi** [24] showed in 1982 that this is true in the special case that the graph is endowed with a linear ordering of the vertices. In such a case it is possible to use tuples of vertices to build a model of a Turing machine inside the graph and imitate the polynomial time property by a suitable fixed point sentence. So in the presence of an ordering the fixed point approach is very powerful. However, what if we do not have an ordering of the vertices of the graph?

It is relatively easy to see with present-day game-theoretic techniques that the polynomial time query "the number of vertices is even" is not expressible in fixed point logic on unordered graphs. The same is true of *all* non-trivial counting queries. This observation has led to the following problem:

$$\text{Is there some natural extension of fixed point lo-} \qquad (4)$$
$$\text{gic which expresses exactly the polynomial time}$$
$$\text{queries on unordered graphs?}$$

raised first by **Chandra** and **Harel** [3].

If it were to be the case that P = NP, then such a natural extension would exist, namely existential second order logic Σ_1^1, defined below. First let us observe that first order logic FO makes sense in a framework that is much more general than just graphs. It is customary in logic to consider structures of a very general type, such as ordered structures, directed graphs, hypergraphs, groups, fields, etc. Common to all these structures is that there is one domain and one or more relations (functions can be treated as relations) and constants on this domain. **Codd** [4] defined on such finite structures the so called *relation algebra* as a kind of minimal database query language. Relation algebra is essentially the same thing as FO. More exactly, a *structure* consists of a set A, a sequence of relations on A, and a sequence of distinguished constants on A. Each relation is a subset of some Cartesian product A^k of A. The number k is called the *arity* of the relation. First order logic for such abstract structures has a name for each relation and constant of the structure. (We use the same symbol for an object and its name, whenever no confusion arises.) The names of the relations and constant of a structure is called the *vocabulary* of the structure.

For example, if (G, E) is a graph and X is a subset of G, we can form a new structure (G, E, X) which has one relation E of arity 2 and one relation X of arity 1. We can say in first order logic that X contains neighbors of its elements (i.e. X is a union of connected components):

$$\forall x \forall y ((X(x) \wedge E(x, y)) \to X(y)).$$

Here $X(t)$ is interpreted as "t is in X". In existential second-order logic Σ_1^1 we can form expressions such as

$$\exists X(X(x) \land \neg X(y) \land \forall z \forall u((X(z) \land E(z,u)) \to X(u))). \tag{5}$$

Here $\exists X$ is a so called second order quantifier because it binds a relation variable X rather than an element variable like $\exists z$ in (3). Note that (5) says that there is no path from x to y. Thus it is equivalent to $\neg\text{Conn}(x,y)$. This shows that Σ_1^1 can express things which are not first order definable.

No-one knows whether Σ_1^1 is closed under negation or not in the framework of finite models. In infinite models infinity itself is a Σ_1^1 concept. Its complement - finiteness - is not Σ_1^1, as follows easily from the so called Compactness Theorem of infinite model theory.

Fagin [9] proved that a query is NP if and only if it is expressible in Σ_1^1, in symbols NP $= \Sigma_1^1$. The role of guessing, inherent in non-deterministic computations, is played by existential second order quantifiers. Fagin's result holds on all structures, ordered or not, because if an order was not present, we could guess an ordering and continue as if the ordering was present. (Recall that ordering is used to simulate a Turing machine inside a structure).

It is possible to express every fixed point query in Σ_1^1. Thus if P $=$ NP, we have the extension Σ_1^1 of FP which expresses exactly the polynomial time queries on graphs or indeed on any structures. The message is, that if we were able to answer question (4) in the negative, we would have proved P $\neq$ NP.

The point of generalized quantifiers (to be defined below) is that they provide a very general yet coherent and mathematically exact approach to extending FO and FP. Proof techniques developed for them provide one possible road to analyzing question (4) and other open questions of descriptive complexity theory.

2   Generalized Quantifiers – Definition

We pointed out above, that there is no fixed point expression (and no first order expression) which would say that the number of vertices of a graph is even. Neither can we say in FP (or FO) that the degree of a vertex is even, or that at last half of the vertices have degree ≥ 3, or that two vertices have the same degree, etc. There is an endless list of examples of simple properties which cannot be captured by FP. This motivates the following idea: We extend FO (and FP) by allowing a new operation $Q_{\text{even}}x(\ldots x \ldots)$ with the interpretation

$$Q_{\text{even}}x(\ldots x \ldots) \Leftrightarrow \text{ the number of } x \text{ with } \ldots x \ldots \text{ is even.} \tag{6}$$

Now the expression $Q_{\text{even}}xE(x,y)$ says in a graph that the vertex y has even degree. By adding another new operation

$$Q_{\text{half}}x(\ldots x \ldots) \Leftrightarrow \text{ at least half of all elements } x \text{ satisfy } \ldots x \ldots,$$

we can say things like at least half of the vertices of a graph have degree ≥ 3:

$$Q_{\text{half}}x\exists y\exists z\exists u(E(x,y) \land E(x,z) \land E(x,u) \land y \neq z \land y \neq u \land z \neq u).$$

The operations Q_{even} and Q_{half} are examples of generalized quantifiers. The extension of FO by Q_{even} is denoted by $\text{FO}(Q_{\text{even}})$ and the extension by Q_{half} is denoted by $\text{FO}(Q_{\text{half}})$. It is easy to verify that all queries that can be expressed in $\text{FO}(Q_{\text{even}})$ or $\text{FO}(Q_{\text{half}})$ are polynomial time computable, but not every polynomial time computable query is expressible in $\text{FO}(Q_{\text{even}})$ or $\text{FO}(Q_{\text{half}})$.

There is an element of arbitrariness in the definition of Q_{even} and Q_{half} above. One gets the feeling that something was needed and it was just thrown in *ad hoc*. The point is that generalized quantifiers provide a way of extending a language in a *minimal* way. Any extension of FO in which we can say one way or other that a predicate is satisfied by an even number of elements, and which satisfies some natural regularity properties, actually contains $\text{FO}(Q_{\text{even}})$.

Let us consider Hamiltonicity of a graph as an example. The probability that a randomly chosen finite graph is Hamiltonian tends to 1 when the size of the graph increases. There are logics, like FO and FP for which a *zero-one law* holds, that is, whatever sentence of the logic we consider, the probability that a randomly chosen finite structure satisfies that sentence tends to 0 or 1 when the size of the graph increases (see [13] for an informal discussion of this). Since Hamiltonicity cannot be expressed either in FO or in FP, the question was raised whether there is some extension of these logics in which Hamiltonicity can be expressed and which has a zero-one law. **Dawar** and **Grädel** [7] proved that the extension of FO obtained by adding the generalized quantifier

$$Q_{\text{Ham}}xy(\ldots x \ldots y \ldots) \Leftrightarrow \text{ the graph with the edge-relation} \atop \{(x,y) : \ldots x \ldots y \ldots\} \text{ is Hamiltonian} \tag{7}$$

does not have the zero-one law. Thus no extension of FO capable of expressing Hamiltonicity can have a zero-one law.

Definition 1 ([21]). *Suppose L is a vocabulary. Any collection Q of structures of vocabulary L, that is closed under isomorphisms, is called a generalized quantifier of vocabulary L.*

The concept of generalized quantifier seems to be extremely general, and does not appear to have anything to do with the ordinary quantifiers $\forall$ and $\exists$.

Let us discuss generality first. The important part of the definition is closure under isomorphisms. To see what this means, and how it limits generality, suppose L consists of one predicate symbol R which is unary i.e. of arity one. Up to isomorphism there are just $n + 1$ non-isomorphic L-structures of size n, namely for each $m \leq n$ the structure $\mathbf{A}_m^n$ with $\{1, \ldots, n\}$ as universe and $\{1, \ldots, m\}$ as the interpretation of R. We can picture these structures as a pyramid (see Figure 1). On vertical rows the size of the model is constant. On rows parallel to the left side of the pyramid the size of the predicate R is constant. On rows parallel to the right side the size of the complement of R is constant.

Now a generalized quantifier of vocabulary L is essentially just a subset of this pyramid. Thus we have a pretty good picture of what kind of generalized quantifiers of vocabulary L there are. For more general vocabularies the picture becomes less and less clear.

$$\mathbf{A}_0^1 \quad \mathbf{A}_1^1$$
$$\mathbf{A}_0^2 \quad \mathbf{A}_1^2 \quad \mathbf{A}_2^2$$
$$\mathbf{A}_0^3 \quad \mathbf{A}_1^3 \quad \mathbf{A}_2^3 \quad \mathbf{A}_3^3$$
$$\mathbf{A}_0^4 \quad \mathbf{A}_1^4 \quad \mathbf{A}_2^4 \quad \mathbf{A}_3^4 \quad \mathbf{A}_4^4$$
$$\ldots$$

Fig. 1. Models with one unary predicate

The relation between generalized quantifiers and ordinary quantifiers becomes evident from the actual use of generalized quantifiers. Suppose Q is as in the above definition. For simplicity, assume L consists of a relation symbol R, which is binary, i.e. of arity two, and of X, which is unary, i.e. of arity one. Using the quantifier Q we can talk about pairs (x, y) of elements satisfying some condition $\ldots x \ldots y \ldots$ and of elements z satisfying another condition $\_\_z\_\_$. The expression

$$Qxyz(\ldots x \ldots y \ldots)(\_\_z\_\_)$$

says in a structure $\mathbf{A}$ with universe A that the L-structure (A, R, X) with universe A and

$$R = \{(x, y) \in A : \ldots x \ldots y \ldots \text{ holds in } \mathbf{A}\}$$
$$X = \{z \in A : \_\_z\_\_ \text{ holds in } \mathbf{A}\},$$

is in the set Q.

In this framework the ordinary quantifier $\exists$ can be defined as a generalized quantifier of vocabulary $\{X\}$, X unary, as follows:

$$\exists = \{(A, X) : X \subseteq A, X \neq \emptyset\}.$$

Thus $\exists x(\ldots x \ldots)$ holds in a structure $\mathbf{A}$ if and only if $(A, X) \in \exists$ where

$$X = \{x \in A : \ldots x \ldots \text{ holds in } \mathbf{A}\}.$$

Similarly

$$\forall = \{(A, X) : X = A\}.$$

Let us denote by $\mathrm{FO}(Q)$ the extension of first order logic by the generalized quantifier Q. There is no special difficulty in adding a generalized quantifier Q to FP, although some attention has to be paid to the details. (Technically speaking, we should speak about the so called inflationary fixed point logic, unless Q is a so called monotone quantifier, but we disregard this detail here.) We denote the result by $\mathrm{FP}(Q)$.

Do generalized quantifiers lead to a solution of question (4)? The answer is "yes and no", as we shall see below. However, it is interesting to note already here that on *almost all* structures PTIME $= \mathrm{FP}(Q_{\mathrm{even}})$ by a recent result of **Hella, Kolaitis and Luosto** [16]. This means that there is a representation of PTIME queries in terms of $\mathrm{FP}(Q_{\mathrm{even}})$ which holds on a randomly chosen finite structure with a probability which tends to one as the size of the structure increases. Even such a weak representation would not be possible with FO or FP alone.

3   A Hierarchy of Generalized Quantifiers

To evaluate the merits of the concept of generalized quantifier we have to look at the results that it gives rise to. For example, hierarchy results are in general rather rare in complexity theory, while strong methods exist for proving hierarchy results for generalized quantifiers.

The hierarchies of generalized quantifiers are based on counting how many variables and in how many formulas the quantifier binds. For example, the quantifier Q_{even} of (6) binds one variable x in one formula. We say that it has type (1). The quantifier

$$Ixy(\ldots x \ldots)(\_\_\_y\_\_\_) \iff \text{there are as many } x \text{ with } \ldots x \ldots$$
$$\text{as there are } y \text{ with } \_\_\_y\_\_\_$$

binds two variables (x and y) in two formulas, one variable per formula. We say that it has type (1,1). Finally Q_{Ham} of (7) binds two variables in one formula, we say that it has type (2).

In general, a *type* is a finite sequence $\tau = (t_1, \ldots, t_k)$ of positive integers with $t_1 \geq \ldots \geq t_k$. A *vocabulary* of type τ has an t_i-ary predicate symbol R_i for each $i = 1, \ldots, k$. A generalized quantifier has type τ if its vocabulary is of type τ.

The type of a vocabulary determines how many models (up to isomorphism) there are of that vocabulary in each size of the domain. If the type τ is *unary*, that is, $t_1 = \ldots = t_k = 1$, then there are

$$\binom{n + 2^k - 1}{2^k - 1}$$

non-isomorphic models of vocabulary L of size n. If the vocabulary L is non-unary, that is, $\max\{t_1, \ldots, t_k\} \geq 2$, then the number of non-isomorphic models of vocabulary L and of size n is harder to compute, but asymptotically it is ([8])

$$\frac{1}{n!} 2^{n^{t_1} + \ldots + n^{t_k}}.$$

These formulas give immediately an idea of the number of quantifiers of a given type. The number *per se* is infinite, but we can get a finite number by considering the restrictions of the quantifier to a fixed finite domain A of cardinality n. If L is unary of type $(t_1, \ldots, t_k)$, the number of restrictions of quantifiers of vocabulary L to A is, of course

$$2^{\binom{n + 2^k - 1}{2^k - 1}}$$

and if L is non-unary, the number is

$$2^{\frac{1}{n!} 2^{n^{t_1} + \ldots + n^{t_k}}}.$$

The point of calculating these numbers is that they give a simple counting method (invented by **Per Lindström**) for proving results about generalized quantifiers. Suppose we want to construct a generalized quantifier Q of type (1,1) which is not definable in $\mathrm{FO}(Q')$ for any generalized quantifier Q' of type (1). Let us make a list

$$\phi_0(Q'), \phi_1(Q'), \phi_2(Q'), \ldots$$

of the possible sentences of $\mathrm{FO}(Q')$ that could give rise to a definition of such a Q. All we have to take care of is that in size n the quantifier Q is defined differently than what $\phi_n(Q')$ says. But (roughly speaking) there are $2^{(n+3)(n+2)(n+1)/3}$ possibilities for Q and only 2^{n+1} possibilities for Q'. So sheer counting shows that such a Q can be constructed. By elaborating this idea it is possible to prove a hierarchy theorem which demonstrates the existence of genuinely new quantifiers on each level of the type hierarchy (see below).

If $\tau = (t_1, \ldots, t_k)$ and $\tau' = (t'_1, \ldots, t'_{k'})$ are types, we let $\tau < \tau'$ if τ precedes τ' in the lexicographic order, that is, $t_i < t'_i$ for the least i such that $t_i \neq t'_i$ (or $k < k'$ if $t_i = t'_i$ for all $i = 1, \ldots, k$). Thus

$$(1) < (1,1) < \ldots < (2) < (2,1) < (2,1,1) < \ldots < (2,2) < \ldots (3) < \ldots.$$

The order-type of this ordering is the infinite ordinal number ω^ω.

Theorem 1 (The Hierarchy Theorem [17]). *For every type $\tau = (t_1, \ldots, t_k)$ there is a generalized quantifier Q of type τ such that Q is not definable in $\mathrm{FO}(Q')$ for any Q' of type $< \tau$.*

In loose terms: we get something genuinely new on every level of types. This result permits many refinements and variations. The most important of them says that Q can be chosen to be LOGSPACE, at least if $t_1 = \ldots = t_k$. Various other constraints can be imposed on Q. An essential feature of this result is that it is a pure existence result, reminiscent of Cantor's diagonalization method. We may force the abstract object Q to satisfy some nice properties but we cannot "put our finger" on Q.

4   Games and Quantifiers

The Hierarchy Theorem of the previous section establishes the richness of the family of all generalized quantifiers. To study properties of individual quantifiers more refined methods are needed. Before introducing the main technical tool in the study of generalized quantifiers, the method of games, let us discuss the general problem of separating models by logical means.

Suppose **A** and **B** are two models of the same vocabulary. We may assume they are both graphs, or we may assume they are two databases (of similar type). Picture 2 shows a simple example of the case that **A** and **B** are graphs. Graph **B** is connected while **A** is not, so the graphs are not isomorphic. What

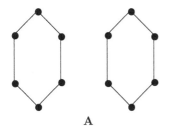 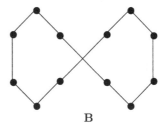

A   B

Fig. 2. Two graphs

is the simplest way of describing the difference between **A** and **B**? Certainly the difference can be described in FO with the sentence

$$\exists x_1 \exists x_2 \ldots \exists x_{12}(E(x_1, x_2) \wedge E(x_2, x_3) \wedge \ldots \wedge E(x_{11}, x_{12}) \wedge E(x_{12}, x_1))$$

but this raises the new question, whether we can do the same with a sentence which would work for larger cycles, too.

On the other hand, the sentence (recall (5))

$$\exists x \exists y(x \neq y \wedge \exists X(X(x) \wedge \neg X(y) \wedge \forall z \forall u((X(z) \wedge E(z, u)) \rightarrow X(u)))) \qquad (8)$$

is a sentence of Σ_1^1 which is true in **A** and false in **B**. To see that this sentence is true in **A**, choose x from the left-hand cycle, y from the right-hand cycle, and let X be the whole left-hand cycle. Note that (8) works regardless of the sizes of **A** and **B** as long as **A** consists of two cycles and **B** of one. So (8) looks like a good logical method to describe the difference between **A** and **B**. But (8) is unsatisfactory as it uses the "second order" quantifier $\exists X$, and no PTIME algorithm is known for checking the truth of sentences involving $\exists X$.

Then there is the fixed point formula $\mathrm{Conn}(x, y)$ of (3), but it is (in a sense) infinitary and therefore unsatisfactory. Of course we can also take the generalized quantifier of type (2)

$$Q_{\mathrm{conn}} xy(\ldots x \ldots y \ldots) \iff \text{the graph which has an edge between}$$
$$\text{every } x \text{ and } y \text{ with } \ldots x \ldots y \ldots$$
$$\text{is connected.}$$

Then $\mathrm{FO}(Q_{\mathrm{conn}})$ can in a trivial way describe a difference between **A** and **B**. The concept of generalized quantifier provides one possible framework for asking questions like: is $\mathrm{FO}(Q_{\mathrm{conn}})$ really the simplest logic in which the difference of **A** and **B** of Picture 2 can be expressed? So what is simpler than $\mathrm{FO}(Q_{\mathrm{conn}})$? What if we replace Q_{conn} by a quantifier of simpler type? The types simpler than that of Q are the unary types (1), (1,1), ...

Proposition 1 ([15]). *No generalized quantifier of a unary type can express connectivity of graphs.*

The main idea behind the proof is the use of the following "bijective" game. The game $BG_k(\mathbf{A}, \mathbf{B})$ is played on two models $\mathbf{A}$ and $\mathbf{B}$. There are two players called I and II. Player II starts by choosing a bijection f_1 between the domain A of $\mathbf{A}$ and the domain B of $\mathbf{B}$. (If there is no such f_1, Player II loses.) Then Player I chooses an element a_1 of A. Now Player II chooses again a bijection f_2 and Player I chooses again an element a_2 of A. The game continues like this for k moves. In the end we have a correspondence

$$
\begin{aligned}
a_1 &\mapsto f_1(a_1) \\
a_2 &\mapsto f_2(a_2) \\
&\vdots \\
a_k &\mapsto f_k(a_k).
\end{aligned}
\tag{9}
$$

Player II is the winner if he has been able to play all the k moves and the produced correspondence (9) is a partial isomorphism, i.e., preserves structure. For example, if $\mathbf{A}$ and $\mathbf{B}$ are graphs, this means that there is an edge between a_i and a_j in $\mathbf{A}$ if and only if there is and edge between $f_i(a_i)$ and $f_j(a_j)$ in $\mathbf{B}$ (and $a_i = a_j \iff f_i(a_i) = f_j(a_j)$). A winning strategy of Player II in $BG_3(\mathbf{A}, \mathbf{B})$ for the graphs of Picture 2 can be described as follows. The choice of f_1 can be arbitrary (because every single element looks the same in both models: edge to the left and edge to the right). After Player I has chosen a_1, Player II chooses f_2 so that it maps any element at a distance $d \leq 2$ from a_1 to an element at the same distance d from $f_1(a_1)$ preserving orientation. Otherwise f_2 can be quite arbitrary as long as it is a bijection. After Player I chooses a_2, Player II chooses f_3 so that it maps a_1, a_2, and their immediate neighbors to $f_1(a_1)$, $f_2(a_2)$, and their immediate neighbors, preserving again orientation. After Player I has chosen a_3 it is easy to see that the mapping $a_i \mapsto f_i(a_i)$ preserves the edge-relation.

Let $\mathbf{Q}_1$ be the family of all unary generalized quantifiers and $\mathrm{FO}(\mathbf{Q}_1)$ the extension of FO by all quantifiers in $\mathbf{Q}_1$.

Theorem 2 ([14]). *A sufficient condition for a property P of finite models to be undefinable in $\mathrm{FO}(\mathbf{Q}_1)$ is that for all numbers k there are models $\mathbf{A}$ and $\mathbf{B}$ such that*

(i) $\mathbf{A}$ has property P,
(ii) $\mathbf{B}$ does not have property P,
(iii) Player II has a winning strategy in the game $BG_k(\mathbf{A}, \mathbf{B})$.

This theorem reduces the logical question whether a property P is expressible in terms of propositional connectives and quantifiers of a very general albeit unary nature, to the purely mathematical question whether certain models with certain combinatorial properties exist. For example, graphs like in Picture 2 together with Theorem 2 provide a proof of Proposition 1.

The proof of Theorem 2 is lengthy but not particularly difficult. The sufficiency of the condition for the undefinability of P is based on the following idea: Let us assume the model $\mathbf{A}$ satisfies a sentence $Qx(\ldots x \ldots)$, supposedly defining P, where Q is a generalized quantifier of type (1). (In general, the defining sentence need not start with Q.) Furthermore, suppose Player II has a winning strategy in $BG_k(\mathbf{A}, \mathbf{B})$ with k sufficiently large. The actual choice of k is based on an inspection of the formula $\ldots x \ldots$. This strategy advises Player II to choose some bijection f_1. Since $Qx(\ldots x \ldots)$ is true in $\mathbf{A}$, the structure (A, X) is in Q, where A is the domain of $\mathbf{A}$ and X is the set of elements x of A which satisfy $\ldots x \ldots$. Let B be the domain of $\mathbf{B}$ and Y the set of elements x of B which satisfy the same condition $\ldots x \ldots$ in $\mathbf{B}$. Now comes the fundamental idea of the game BG_k. We claim that

$$f_1 : (A, X) \cong (B, Y). \tag{10}$$

Suppose $x \in A$. We have to prove $a \in X \iff f_1(a) \in Y$. Suppose therefore e.g. $x \in X$. Thus x satisfies the condition $\ldots x \ldots$ in $\mathbf{A}$. Let us let Player I choose $a_1 = x$ in the game $BG_k(\mathbf{A}, \mathbf{B})$. Because Player II is playing a winning strategy, we know that a_1 satisfies the same basic relations of $\mathbf{A}$ as $f_1(a_1)$ satisfies in $\mathbf{B}$. By formulating the induction hypothesis appropriately, we actually know that a_1 satisfies the same "definable" relations in $\mathbf{A}$ as $f_1(a_1)$ satisfies in $\mathbf{B}$. Here "definable" means definability in a restricted sense, which however includes the relation $\ldots x \ldots$. Since we know that a_1 satisfies condition $\ldots a_1 \ldots$ in $\mathbf{A}$, we can conclude that $f_1(a_1)$ satisfies condition $\ldots f_1(a_1) \ldots$ in $\mathbf{B}$. In particular, $f_1(a_1) \in Y$, as desired.

Now that we know (10), we may draw from $(A, X) \in Q$ the conclusion $(B, Y) \in Q$, which means, by definition, that the sentence $Qx(\ldots x \ldots)$ is true in $\mathbf{B}$. Thus $Qx(\ldots x \ldots)$ cannot, after all, define the property P.

Theorem 2 has been successfully used to show that various graph properties (e.g. planarity) are not expressible in terms of unary quantifiers [23]. Such results can be seen as formalizations of the vague intuition that some properties of binary relations cannot be reduced to properties of cardinalities of definable sets.

What is really remarkable about Theorem 2 is that it generalizes both to non-unary quantifiers and to extensions of *fixed point logic* by generalized quantifiers. Let us call a type $\tau = (t_1, \ldots, t_k)$ *r-ary* if $\max\{t_1, \ldots, t_k\} \leq r$. A vocabulary (and a generalized quantifier) is called *r-ary* if its type is r-ary. The word "binary" is generally used for 2-ary.

The game $BG_k^r(\mathbf{A}, \mathbf{B})$, an r-ary version of $BG_k(\mathbf{A}, \mathbf{B})$, is defined like the game $BG_k(\mathbf{A}, \mathbf{B})$ except that when I moves, he chooses an r-tuple $(a_i^1, \ldots, a_i^r)$ rather than a single element a_i. So after k moves we have the correspondence

$$
\begin{aligned}
a_1^j &\mapsto f_1(a_1^j), j = 1, \ldots, r \\
&\;\;\vdots \\
a_k^j &\mapsto f_k(a_k^j), j = 1, \ldots, r
\end{aligned}
\tag{11}
$$

and Player II wins if this is a partial isomorphism. This game is immensely more difficult to win for Player II than $BG_k(\mathbf{A}, \mathbf{B})$ since already f_1 has to preserve all r-tuples. For graphs this means that Player II cannot even make the first move without losing unless $\mathbf{A}$ and $\mathbf{B}$ are isomorphic.

Let $\mathbf{Q}_r$ denote the family of all r-ary generalized quantifiers, and $FO(\mathbf{Q}_r)$ the corresponding extension of FO.

Theorem 3 ([14]). *Theorem 2 holds for r-ary generalized quantifiers, that is, if $\mathbf{Q}_1$ is replaced by $\mathbf{Q}_r$ and $BG_k(\mathbf{A},\mathbf{B})$ by $BG_k^r(\mathbf{A},\mathbf{B})$.*

By means of this criterion it is possible to show that the following "Ramsey-quantifier" of type (r)

$$Ram^r\mathbf{x}(\ldots\mathbf{x}\ldots) \iff \exists X(X \text{ large } \wedge \forall\mathbf{x} \in [X]^r(\ldots\mathbf{x}\ldots)) \qquad (12)$$

is not definable in $FO(\mathbf{Q}_{r-1})$[18]. Here "large" can be anything reasonable, e.g. $n/2$ or $\log(n)$, where n is the size of the model. We have denoted an r-tuple $(x_1,\ldots,x_r)$ by the bold face $\mathbf{x}$. $[X]^r$ means the set of all r-tuples of distinct elements of X. The Ramsey-quantifier is not known to be PTIME in general. Indeed, in many natural instances it is NP-complete. But there are other explicit r-ary quantifiers, even in FP, which are not definable in $FO(\mathbf{Q}_{r-1})$. One example is the transitive closure quantifier on r-tuples (see below) [10]. An interesting conclusion from this is:

Theorem 4 ([14]). *Fixed point logic FP cannot be represented as the extension of FO by finitely many generalized quantifiers.*

Namely, the arities of the finitely many quantifiers would have a common upper bound r, so such a representation would contradict the existence of FP queries, which are not definable in $FO(\mathbf{Q}_r)$. This is a strong manifestation of the inherent incapability of (finitely many) generalized quantifiers to express recursion.

What about *combining* recursion and generalized quantifiers? We have already remarked that any generalized quantifier can be added to fixed point logic FP. Let $FP(\mathbf{Q}_r)$ denote the extension of FP by all r-ary quantifiers.

We can replace the logic $FO(\mathbf{Q}_1)$ by the logic $FP(\mathbf{Q}_r)$ in Theorem 2, if we simultaneously replace the game $BG_k(\mathbf{A},\mathbf{B})$ by a new game $BPG_k^r(\mathbf{A},\mathbf{B})$, called the *bijective pebble game*, which we now define. This game is like $BG_k^r(\mathbf{A},\mathbf{B})$ but it is potentially infinitely long (although it does not make sense to play more than n^k moves, where n is the size of $\mathbf{A}$). Therefore it is even harder for Player II than $BG_k^r(\mathbf{A},\mathbf{B})$. To counterbalance the length there is a special mechanism to limit the size of the final correspondence (cf. (11)). There are k pebbles. While in $BG_k^r(\mathbf{A},\mathbf{B})$ Player I chooses some r-tuple, in the game $BPG_k^r(\mathbf{A},\mathbf{B})$ he puts pebbles on elements of an r-tuple. What is the difference? Eventually he runs out of pebbles! Then he can take back some pebbles that he has used already and reuse them. At some point Player I decides that the game has lasted long enough. Now we look at the correspondence generated by the elements that have a pebble on them in the final position. So we form the mapping

$$a_{i_1}^{j_1} \mapsto f_{i_1}(a_{i_1}^{j_1})$$
$$\vdots \qquad\qquad (13)$$
$$a_{i_k}^{j_k} \mapsto f_{i_k}(a_{i_k}^{j_k}),$$

where $a_{i_1}^{j_1}, \ldots, a_{i_k}^{j_k}$ is the sequence of elements with a pebble on them at the end. If this is a partial isomorphism, then Player II has won.

The challenge that $BPG_k^r(\mathbf{A}, \mathbf{B})$ presents for Player II is that he should find a strategy which works no matter how long the game has been going on and no matter how long it will go on. For example, if $\mathbf{A}$ and $\mathbf{B}$ are as in Picture 2, Player II wins $BPG_k^1(\mathbf{A}, \mathbf{B})$ if and only if $k = 1$.

Theorem 5 ([14]). *Theorem 2 holds for fixed point logic and r-ary generalized quantifier, that is, if* $\mathrm{FO}(\mathbf{Q}_1)$ *is replaced by* $\mathrm{FP}(\mathbf{Q}_r)$ *and* $BG_k(\mathbf{A}, \mathbf{B})$ *by* $BPG_k^r(\mathbf{A}, \mathbf{B})$.

With this criterion it is possible to prove that certain properties of r-ary relations are not expressible in terms of $(r-1)$-ary generalized quantifiers even if taking least fixed points of formulas are allowed. The first result in this direction, due to **Cai**, **Fürer** and **Immerman** [2], exhibited a LOGSPACE property of graphs that is not expressible in $\mathrm{FP}(\mathbf{Q}_1)$. This demonstrates in a powerful way the impossibility of solving question (4) by means of fixed points and quantifiers which merely count sizes of definable sets. A further result of **Hella** [14] showed that for every r there are LOGSPACE properties of finite models that are not expressible in $\mathrm{FP}(\mathbf{Q}_r)$. Hence:

Theorem 6 ([14]). *On unordered finite models, PTIME is not the extension of fixed point logic by finitely many generalized quantifiers.*

This shows that if we want to answer (4) affirmatively, we have to look beyond fixed point logic and finite collections of generalized quantifiers.

The difference between the Hierarchy Theorem (Theorem 1) and the results of this section is that the counting method of the Hierarchy Theorem simply gives the existence of a quantifier with certain properties, with no concern to whether the quantifier has any intuitive meaning. The more elaborate game-theoretic methods of this section make it possible to take a concrete meaningful quantifier, like the Ramsey-quantifier, and prove that it cannot be reduced to simpler quantifiers.

5   Quantifier Schemata

We have pointed out that PTIME can be expressed in logical formalism as fixed point logic, provided that we restrict ourselves to ordered models. We have also pointed out that there is an extension of fixed point logic by a single generalized quantifier, which captures PTIME on almost all unordered finite models. Finally, we have concluded that in the framework of all unordered finite models there is no extension of fixed point logic by finitely many generalized quantifiers that would give all of PTIME.

Can we capture PTIME by adding an *infinite* number of generalized quantifiers to fixed point logic? There is a trivial answer. We can take one new quantifier for each PTIME query and add the resulting infinitely many quantifiers to FO.

Surely we get PTIME, but we have not gained anything. It makes more sense to look at infinite collections of quantifiers arising from some effective process, and then ask, can we get all of PTIME.

There are several ways in which a single generalized quantifier can give rise to an infinite sequence of quantifiers. Recall the Ramsey-quantifier Ram^r defined in (12). We can think of it as arising from the much simpler quantifier

$$Q_f x(\ldots x \ldots) \iff \exists X \left(|X| \geq f(n) \land \forall x \in X(\ldots x \ldots) \right),$$

where $f : \mathbf{N} \to \mathbf{N}$ and n is the size of the model. With this quantifier one can say (by choosing $f(n) = \lfloor n/2 \rfloor + 1$) things like:

"Most vertices have a green neighbor."

Let us denote Ram^r by $Ram^r(Q_f)$ if "X large" in (12) is defined as $|X| \geq f(n)$, where again n denotes the size of the model. $Ram^r(Q_f)$ is called a *Ramsey lift* of Q_f. So with the sentence

$$Ram^2(Q_f)xyE(x,y)$$

we can say

"Most vertices are neighbors of each other."   (14)

For every choice of f, the quantifier Q_f of type (1) gives rise to the infinite sequence of quantifiers $Ram^r(Q_f)$ of higher and higher type. It turns out that for non-trivial f the quantifier $Ram^n(Q_f)$ cannot be defined in terms of quantifiers of smaller type even if fixed points are used. But we would not get all of PTIME even if we added all possible Ramsey lifts of quantifiers of type (1) [17]. Other lifts of the nature of the Ramsey lift have been considered, especially in the study of natural language semantics, where they are used to formalize expressions like

Most boys in my class and most girls in your class have all dated each other.

However, the most interesting lift from the point of view of descriptive complexity theory is the *resumption* or *vectorization* lift. In a few words, the vectorization of a quantifier Q says about tuples what Q itself says about elements. Thus the second vectorization of Q_{even} says "for an even number of pairs (x, y) we have $\ldots x \ldots y \ldots$. In general, let Q be a quantifier of type $\tau = (t_1, \ldots, t_k)$. Then its m'th vectorization $Res^m(Q)$ is a quantifier of type $(mt_1, \ldots, mt_k)$. So the arity of the quantifier increases by a factor of m. A quantifier Q of type τ is a class of models of vocabulary L, where L consists of a t_i-ary relation symbol $R_i(x_1, \ldots, x_{t_i})$ for each $i = 1, \ldots, k$. To define $Res^m(Q)$ we use the vector notation $\mathbf{z}$ to denote a sequence $(z_1, \ldots, z_m)$ of m variables. Let the language L' consist of an (mt_i)-ary relation symbol $R_i'(x_1, \ldots, x_{mt_i})$ for each $i = 1, \ldots, k$. Then $Res^m(Q)$ is the class of models $(A, R_1', \ldots, R_k')$ of vocabulary L' for which the L-structure $(A^m, R_1, \ldots, R_k)$ is in Q, where

$$R_i = \{((a_1^1, \ldots, a_{t_i}^1), \ldots, (a_1^m, \ldots, a_{t_i}^m)) \in (A^m)^{t_i} :$$
$$R_i'(a_1^1, \ldots, a_{t_i}^1, \ldots, a_1^m, \ldots, a_{t_i}^m)\}.$$

For example, suppose Q is the type $(2,1)$ quantifier which consists of structures (A, E, U), where $U \subseteq A$ and (A, E) is a graph with a clique of the size of the set U. So in a model with domain A the expression

$$Qxyz(\ldots x \ldots y \ldots)(\_\_\_z\_\_\_)$$

says that $(A, \{(a, b) \in A^2 : \ldots a \ldots b \ldots\})$ is a graph with a clique with as many elements as there are $a \in A$ with $\_\_\_a\_\_\_$, while $Res^m(Q)$ is of type $(2m, m)$ and

$$Q\mathbf{xyz}(\ldots \mathbf{x} \ldots \mathbf{y} \ldots)(\_\_\_\mathbf{z}\_\_\_)$$

says in the same model that $(A^m, \{(\mathbf{a}, \mathbf{b}) \in (A^m)^2 : \ldots \mathbf{a} \ldots \mathbf{b} \ldots\})$ is a graph with a clique with as many elements as there are sequences $\mathbf{a} \in A^m$ with $\_\_\_\mathbf{a}\_\_\_$.

Note that if Q is PTIME, then so is every $Res^m(Q)$, so the vectorization lift is computationally simpler than the Ramsey lift. We denote by $FO(Q^{<\omega})$ the extension of FO by all the vectorizations $Res^m(Q)$, $m = 1, 2, \ldots$, of Q.

As an important concrete example, consider the *Transitive Closure* quantifier of type $(2, 1, 1)$:

$$TC = \{(A, E, X, Y) : (A, E) \text{ is a graph}, X \subseteq A, Y \subseteq A$$
$$\text{and from every } x \in X \text{ there is a path in the graph}$$
$$\text{to some } y \in Y.\}$$

Theorem 7 ([20]). *NLOGSPACE= $FO(TC^{<\omega})$ on ordered models.*

It is interesting to note that before this result it was not even known whether NLOGSPACE is closed under complements. It is an open problem, whether there is some natural logic L such that NLOGSPACE= $FO(L)$ holds on unordered models.

The *Alternating Transitive Closure* quantifier ATC consists of models of the form (A, E, X, Y), where (A, E) is a graph, $X \subseteq A, Y \subseteq A$, and every $x_0 \in X$ has a neighbor x_1 whose every neighbor x_2 has a neighbor x_3 whose every neighbor x_4 has a neighbor x_5 ...etc... until we come to an element of Y. Immerman [20] proved that PTIME= $FO(ATC^{<\omega})$ on ordered finite models. This result has the following interesting version on *all* models, ordered or unordered:

Theorem 8 ([5]). $FP = FO(ATC^{<\omega})$.

So although no finite sequence of generalized quantifiers can capture all of FP on unordered models (Theorem 4), the infinite sequence $ATC^{<\omega}$ is capable of the job.

Generalized quantifiers are thus, after all, able to express recursive definitions, as soon as sufficient arities, i.e. sufficiently long tuples are available. Still, it remains an open problem, whether PTIME $= FO(Q^{<\omega})$ for some Q on unordered models. But Dawar has proved the following interesting result, which shows that the approach of (vectorizations) of generalized quantifiers is at least as good as any other:

Theorem 9 ([6]). *If question (4) has a positive answer (in an exact sense), then there is a generalized quantifier Q so that $PTIME = \mathrm{FO}(Q^{<\omega})$ on all finite models.*

The methods available at the moment in the study of generalized quantifiers can be effectively used to study definability questions concerning individual quantifiers and families of quantifiers with a bound on arities. Unfortunately the same methods become extremely hard when applied to quantifier schemata.

References

1. A.V. Aho and J.D. Ullman, Universality of data retrieval languages, *Sixth ACM Symposium on Principles of Programming Languages*, 1979, 110–117.
2. J. Cai, M. Fürer and N. Immerman, An Optimal Lower Bound on the Number of Variables for Graph Identification, *Combinatorica* 12:4 (1992), 389–410.
3. A. Chandra and D. Harel. Structure and complexity of relational queries. *Journal of Computer and System Sciences*, 25:99–128, 1982.
4. E.F. Codd, Relational completeness of database sublanguages, in: *Database Systems* (R.Rustin, ed.), Prentice-Hall, 1972, 65–98.
5. E. Dahlhaus, Skolem normal forms concerning the least fixpoint, in: *Computation theory and logic* (E. Börger, ed.), 101–106, Lecture Notes in Comput. Sci., 270, Springer, Berlin-New York, 1987.
6. A. Dawar, Generalized quantifiers and logical reducibilities, *Journal of Logic and Computation*, 5(1995), 213–226.
7. A. Dawar and E. Grädel, Generalized quantifiers and 0-1 laws, *Proc. 10th IEEE Symp. on Logic in Computer Science*, 1995, 54–64.
8. R. Fagin, The number of finite relational structures, *Discrete Mathematics* 19(1977), 17–21.
9. R. Fagin, Generalized first-order spectra and polynomial-time recognizable sets, in: *Complexity of Computation* (R. Karp, ed.) SIAM-AMS Proc. 7, 1974, 27–41.
10. M. Grohe and L. Hella, A double arity hierarchy theorem for transitive closure logic, *Archive for Mathematical Logic*, 35(3): 157-172, 1996.
11. Y. Gurevich. Toward logic tailored for computational complexity. In M. M. Richter et al., editor, *Computation and Proof Theory, Lecture Notes in Mathematics 1104*, pages 175–216. Springer-Verlag, 1984.
12. Y. Gurevich. Logic and the challenge of computer science. In E. Börger, editor, *Current trends in theoretical computer science*, pages 1–57. Computer Science Press, 1988.
13. Y. Gurevich, Zero-one laws, *EATCS Bulletin* 46 (1992), 90-106.
14. L. Hella, Logical hierarchies in PTIME, *Information and Computation* 129: 1-19, 1996.
15. L. Hella and G. Sandu, Partially ordered connectives and finite graphs, in: *Quantifiers: Logics, models and computation* (M. Krynicki, M. Mostowski and L. Szczerba, eds.), vol II, Kluwer Academic Publishers 1995, 79–88.
16. L. Hella, Ph. Kolaitis and K. Luosto, Almost everywhere equivalence of logics in finite model theory, *Bulletin of Symbolic Logic* 2(4), 1996, 422-443.
17. L. Hella, K. Luosto and J. Väänänen, The hierarchy theorem for generalized quantifiers. *Journal Symbolic Logic* 61 (1996), no. 3, 802–817.

18. L. Hella, J. Väänänen and D. Westerståhl, Definability of polyadic lifts of generalized quantifiers, *Journal of Logic, Language and Information* 6 (1997), no. 3, 305-335.

19. N. Immerman, Relational queries computable in polynomial time, *Information and Control* 68 (1986), 86–104.

20. N. Immerman, Languages that capture complexity classes, *SIAM J. Comput.* 16, No. 4 (1987), 760-778.

21. P. Lindström, First order predicate logic with generalized quantifiers, *Theoria* 32 (1966), 186–195.

22. A. Mostowski, On a generalization of quantifiers. *Fundamenta Mathematicae* 44 (1957) 12–36.

23. J. Nurmonen, On winning strategies with unary quantifiers, *Journal of Logic and Computation*, 6(6): 779-798, 1996.

24. M. Vardi, Complexity of relational query languages, *14th ACM STOC Symposium* (1982), 137–146.

Counting and Locality over Finite Structures
A Survey

Leonid Libkin[1] and Juha Nurmonen[*2]

[1] Bell Laboratories, 600 Mountain Avenue, Murray Hill, NJ 07974, USA
libkin@bell-labs.com
[2] Department of Mathematics and Computer Science, University of Leicester,
University Road, Leicester LE1 7RH, UK
j.nurmonen@mcs.le.ac.uk

Abstract. We survey recent results on logics with counting and their local properties. We first consider game-theoretic characterizations of first-order logic and its counting extensions provided by unary generalized quantifiers. We then study Gaifman's and Hanf's locality theorems, their connection with game characterizations, and examples of their usage in proving expressivity bounds for first-order logic and its extensions. We review the abstract notions of Gaifman's and Hanf's locality, and show how they are related. We also consider a closely related bounded degree property, and demonstrate its usefulness in proving expressivity bounds. We discuss two applications. One deals with proving lower bounds for the complexity class TC^0. In particular, we use logical characterization of TC^0 and locality theorems for first-order with counting quantifiers to provide lower bounds. We then explain how the notions of locality are used in database theory to prove that extensions of relational calculus with aggregate functions and grouping still lack the power to express fixpoint computation.

1 Introduction

Finite model theory is an active area of research, mostly due to its connections to theoretical computer science, in particular, database theory [1] and complexity theory [13,28]. Several important complexity classes have nice logical characterizations. For instance, the star-free languages are exactly the ones definable in first-order logic FO, least fixpoint logic LFP captures PTIME, and partial fixpoint logic PFP captures PSPACE on ordered finite structures (see [13]). These logics also have their counterparts in the theory of database query languages. For example, relational calculus, that underlies majority of practical query languages, has exactly the power of first-order logic. The language Datalog with negation corresponds to LFP, and the extension of relational calculus with while loops corresponds to PFP (see [1]).

Several counting properties have also been considered in complexity and database theory. We give a few examples here. A logic can be extended with a

---

[*] Supported by EPSRC grant GR/K 96564.

linearly ordered second sort universe of numbers, and *counting quantifiers* 'there are at least n elements.' A very strong result in [9] shows that least fixpoint logic with such second sort counting fails to capture the complexity class PTIME. In circuit complexity theory, *counting modulo quantifiers* have been studied. Allowing gates which count inputs modulo a constant p, for every p, in the definition of AC^0, one obtains the class ACC. Again, this class has a logical characterization, and the problem whether the containment $ACC \subseteq NC^1$ is strict, is still open. In [5], it was shown that the class TC^0, that extends AC^0 with threshold gates, can be captured on ordered structures by first-order logic with the second sort counting, and simple arithmetic predicates on numbers.

In database theory, one often extends traditional first-order based languages with aggregate functions, such as summing up all values in a column in a relation, or finding the average value. One also extends those languages with grouping, that permits queries such as finding the average salary in each department. While the expressive power of relational calculus and many other query languages is well understood, much less is known about their aggregate extensions.

These applications have motivated a systematic study of the expressive power of counting properties from the viewpoint of finite model theory. In finite model theory, one normally uses games to prove expressivity bounds. For example, Ehrenfeucht-Fraïssé games are used for first-order logic, and bijective Ehrenfeucht-Fraïssé games are used for FO extended with unary generalized quantifiers. In Section 3 we review rules of such games for FO and some of its extensions with counting.

Playing a game often involves a complicated combinatorial argument. Many results on the expressive power of first-order logic and its extensions with counting, in particular, those obtained as an attempt to avoid game arguments, give us the intuition that these logics can express only local properties, and lack a mechanism for fixpoint computation. In Section 4, we review several results of this kind. We consider Gaifman's theorem [18] for first-order logic, which shows that every first-order formula is equivalent to a local one, in the sense that only a small part of a structure is relevant for evaluating the query given by a formula. We also study modifications of Hanf's result [22]. In this approach one counts the number of isomorphism types of fixed radius neighborhoods of points. If the result of this counting satisfies certain criteria, then the structures considered are guaranteed to be elementary equivalent in a certain logic. This technique has been modified for first-order logic [17], first-order logic with counting modulo quantifiers [41] and first-order logic extended by all unary generalized quantifiers [40], for the case of finite structures. Proofs of applicability of Hanf's technique typically are not very difficult [17,15,40,42]. We will see some examples in Section 4.

The above results have motivated a study of general notions of locality [32,24]. We review this line of work in Section 5. We show that Gaifman's theorem gives rise to two general notions, one for sentences and one for open formulas. We formulate an abstract notion of locality that captures Hanf's condition, and study the relationship between the notions of locality. We also consider the bounded degree property, which is implied by all other notions of locality, and is particularly well suited for proving expressivity bounds, especially for properties involving fixpoint computation.

In Section 6, we discuss applications of the concept of locality in complexity theory. In particular, we study the relationship between the circuit complexity class TC^0 and other complexity classes such as L and NL. In Section 7, we consider applications in database theory. We review results on expressive power of relational languages that resemble commercial languages such as SQL. We show that queries such as transitive closure, are inexpressible in a theoretical language that has the power of core SQL.

2   Preliminaries

A relational signature σ is a finite set of relation symbols $\{R_1, ..., R_l\}$, each of which has an arity $p_i > 0$. We write σ_n for σ extended with n new constant symbols. A σ-structure is $\mathcal{A} = (A, \overline{R}_1, \ldots, \overline{R}_l)$, where A is a non-empty set and $\overline{R}_i \subseteq A^{p_i}$ interprets R_i. If the *universe* A is finite, the structure $\mathcal{A}$ is called finite. Unless mentioned otherwise, all structures considered here are assumed to be finite. When the notation is clear from the context, we write R_i in place of $\overline{R}_i$. The class of finite σ-structures is denoted by $\text{STRUCT}[\sigma]$. For instance, a graph $\mathcal{A} = (A, E)$ is a structure over a signature which consists of a single binary relation symbol E. If E is required to be symmetric, then $\mathcal{A}$ is an undirected graph.

If $X \subseteq A$, by $\mathcal{A} \restriction X$ we mean the structure with universe X, where the interpretation of each R_i is restricted to X. An *isomorphism* $f : \mathcal{A} \to \mathcal{B}$ is a bijection $A \to B$ such that $(a_1, \ldots, a_{p_i}) \in \overline{R}_i$ holds in $\mathcal{A}$ if and only if $(f(a_1), \ldots, f(a_{p_i})) \in \overline{R}_i$ holds in $\mathcal{B}$. If $a_1, \ldots, a_n \in A$ and $b_1, \ldots, b_n \in B$, we say that $\{(a_i, b_i) \mid 1 \leq i \leq n\}$ is a *partial isomorphism* $\mathcal{A} \to \mathcal{B}$ if it is an isomorphism $\mathcal{A} \restriction \{a_1, \ldots, a_n\} \to \mathcal{B} \restriction \{b_1, \ldots, b_n\}$.

Every formula $\psi(x_1, \ldots, x_m)$ with free variables $x_1, \ldots, x_m$ defines a *query* which maps a σ-structure $\mathcal{A}$ to an m-ary relation $q_\psi(\mathcal{A}) = \{\mathbf{a} \in A^m \mid \mathcal{A} \models \psi(\mathbf{a})\}$. We denote the corresponding structure by $\psi[\mathcal{A}] = (A, q_\psi(\mathcal{A}))$. An m-ary query q is definable in a logic $\mathcal{L}$ if there is a formula $\varphi(\mathbf{x})$ of that logic such that for every $\mathcal{A} \in \text{STRUCT}[\sigma]$ we have $q(\mathcal{A}) = \{\mathbf{a} \in A^m \mid \mathcal{A} \models \varphi(\mathbf{a})\}$. As an example, consider the *transitive closure* query *TRCL*. Suppose we are given a graph $\mathcal{A} = (A, E)$. Then *TRCL* consists of all pairs $(a, b) \in A^2$ such that there is an E-path from a to b, that is,

$$q_{TRCL} = \left\{ (a, b) \in A^2 \,\middle|\, \begin{array}{l} \exists a_1, \ldots, a_k \in A \text{ such that } a = a_1 \text{ and } b = a_k \text{ and} \\ E(a_i, a_{i+1}) \text{ for all } i \end{array} \right\}.$$

This query is definable in many fixpoint logics, and in second-order logic, but we shall see that it is not definable in FO and its counting extensions.

The *Gaifman graph* $\mathcal{G}(\mathcal{A})$ of a σ-structure $\mathcal{A}$ is the undirected graph (A, E) where $(a, b) \in E$ if and only if there is a tuple $\mathbf{t} \in \overline{R}_i$ for some $R_i \in \sigma$ such that $a, b \in \mathbf{t}$. Note that if $\mathcal{A}$ is an undirected graph, then $\mathcal{A} = \mathcal{G}(\mathcal{A})$. The *degree* of a point is its degree in the Gaifman graph. The *distance* $d(a, b)$ is the length of the shortest path from a to b in $\mathcal{G}(\mathcal{A})$. For $a \in A$, its *r-sphere* is $S_r^{\mathcal{A}}(a) = \{b \in A \mid d(b, a) \leq r\}$. For an n-tuple t we define $S_r^{\mathcal{A}}(t) = \cup_{a \in t} S_r^{\mathcal{A}}(a)$. The *r-neighborhood* of a tuple t is the σ_n-structure $N_r^{\mathcal{A}}(t) = (\mathcal{A} \upharpoonright S_r^{\mathcal{A}}(t), t)$. That is, it is the restriction of $\mathcal{A}$ to $S_r^{\mathcal{A}}(t)$, with t being n distinguished constants. For instance, if $\mathcal{A}$ is a chordless cycle of length at least $2r + 2$ and $a \in A$, then $N_r^{\mathcal{A}}(a)$ is the chordless path of length $2r + 1$ with one distinguished point a. We denote the *isomorphism type* of $N_r^{\mathcal{A}}(t)$ by $tp_r^{\mathcal{A}}(t)$. We emphasize here that if $h : N_r^{\mathcal{A}}(a) \to N_r^{\mathcal{A}}(b)$ is an isomorphism, where $a = (a_1, \ldots, a_n)$ and $b = (b_1, \ldots, b_n)$, then $h(a_1) = b_1, \ldots, h(a_n) = b_n$, as neighborhoods $N_r^{\mathcal{A}}(a)$ and $N_r^{\mathcal{A}}(b)$ are σ_n-structures.

If the structure $\mathcal{A}$ is understood, we omit it from the notations when convenient.

3   Logics and Games

In this section we introduce the logics that are considered in this paper. These logics are first-order logic FO and its extensions with various generalized quantifiers. All logics we consider are closed under Boolean connectives and first-order quantification, and are regular in the sense of [12,31] (we do not go into details in this survey). We also review game-theoretic characterizations for elementary equivalence in these logics, and show how the games can be used to prove expressivity bounds.

We use the standard definitions for formulas and semantics of first-order logic FO (see e.g. [13]). Equality is treated as a special relation symbol which is not a member of the signature. The *quantifier rank* of a formula φ is defined to be the depth of quantifier nesting in φ, and is denoted by $qr(\varphi)$.

The rules of the *first-order Ehrenfeucht-Fraïssé game* are as follows. There are two players, *the spoiler* and *the duplicator*. Two σ-structures $\mathcal{A}$ and $\mathcal{B}$ and the number of rounds, say n, are given. In each round the spoiler first selects a point of one of the structures and the duplicator selects a point of the other structure. Let $a_1, \ldots, a_n$ and $b_1, \ldots, b_n$ be the points selected after the last round from $\mathcal{A}$ and $\mathcal{B}$, respectively. The duplicator is declared the winner if $\{(a_i, b_i) \mid 1 \leq i \leq n\}$ is a partial isomorphism $\mathcal{A} \to \mathcal{B}$; otherwise the spoiler wins. We say that a player has *a winning strategy* if he can guarantee a win, no matter how the other player plays. This game is interesting because of the following result.

Theorem 1 (cf. [13]). *Let* $\mathcal{A}, \mathcal{B} \in \text{STRUCT}[\sigma]$. *Then the duplicator has a winning strategy in the n-round Ehrenfeucht-Fraïssé game if and only if $\mathcal{A}$ and $\mathcal{B}$ agree on all first-order sentences of quantifier rank up to n.*   □

We use the notation $\mathcal{A} \equiv_{\text{FO}}^n \mathcal{B}$ if the duplicator has a winning strategy in the n-round game on $\mathcal{A}$ and $\mathcal{B}$. The above theorem can be used to provide the following tool for proving expressivity bounds.

Corollary 1 (cf. [13]). *A class* $\mathcal{C} \subseteq \mathrm{STRUCT}[\sigma]$ *is not definable in first-order logic if and only if for every n there are σ-structures $\mathcal{A} \in \mathcal{C}$ and $\mathcal{B} \notin \mathcal{C}$ such that the duplicator has a winning strategy in the n-round Ehrenfeucht-Fraïssé game on $\mathcal{A}$ and $\mathcal{B}$.*   □

In other words, if for every n we can find one σ-structure from the class $\mathcal{C}$ and another from the complement of $\mathcal{C}$, such that the duplicator can maintain a partial isomorphism for n rounds, then $\mathcal{C}$ is not definable in FO.

We now give some examples. Note that in those examples, we do not spell out every single detail of the game argument – this may require more space than this entire section. We shall offer much simpler proofs of the existence of winning strategies in the next section, after we have introduced the ideas of locality.

First, one can use Ehrenfeucht-Fraïssé games to show that connectivity of finite graphs cannot be expressed in first-order logic. Assume that connectivity is definable, and take as a counterexample, for each finite n, $\mathcal{A}$ to be a chordless cycle of length 2^n and $\mathcal{B}$ to be a disjoint union of two chordless cycles of length 2^n. The duplicator's strategy in the n-round Ehrenfeucht-Fraïssé game is to preserve, in each round $j \leq n$, distances up to 2^{n-j}. The only way the spoiler could win is to show that in $\mathcal{B}$ there are two points with no path between them, whereas in $\mathcal{A}$ there always is a path between any two points. But if the spoiler cannot build a path between two points in n rounds, it does not matter how far these points are, or whether there is a path between them at all. This informal reasoning can be formalized to show that the duplicator has a winning strategy in the n-round Ehrenfeucht-Fraïssé game over $\mathcal{A}$ and $\mathcal{B}$. This result holds also for ordered structures, see [13]. Note that this shows that the transitive closure query *TRCL* cannot be defined in FO. Indeed, assume that a first-order formula $\psi(x, y)$ defines *TRCL*. Then $\forall x \forall y \psi(x, y)$ would be a first-order sentence defining connectivity.

Next, assume that we are given two distinguished points a and b of a graph $\mathcal{A}$. Then we cannot define, in first-order logic, the property that a and b have equally many neighbors. As a counterexample we can take, for every n, a to be a point which has $n + 2$ neighbors and b to be a point with $n + 1$ neighbors in some graph $\mathcal{A}$. Since there are only n rounds available, the spoiler cannot demonstrate that b has fewer neighbors.

3.1   Unary Quantifiers

A commonly used way to increase the expressive power of first-order logic is to extend it with *generalized quantifiers*, cf. [31,48]. The basic idea is that we are given a class of structures, and we can check whether a substructure defined by a given family of formulas belongs to this class. In computational complexity theory, generalized quantifiers are often considered as oracles. In this section, however, we concentrate more on logical aspects of generalized quantifiers. We now review this method in detail in the case of *unary* generalized quantifiers.

Let σ_k^{unary} be a signature of k unary symbols. Suppose $\mathcal{K}$ is an isomorphism closed class of σ_k^{unary}-structures. Then $\text{FO}(Q_\mathcal{K})$ is the extension of FO by a new formula formation rule:

if $\psi_i(x_i, \boldsymbol{y}_i)$ is a formula of $\text{FO}(Q_\mathcal{K})$ for $i = 1, \ldots, k$, then
$Q_\mathcal{K} x_1, \ldots, x_k(\psi_1(x_1, \boldsymbol{y}_1), \ldots, \psi_k(x_k, \boldsymbol{y}_k))$ is a formula of $\text{FO}(Q_\mathcal{K})$.

The corresponding semantic rule is:

$\mathcal{A} \models Q_\mathcal{K} x_1 \ldots x_k(\psi_1(x_1, \boldsymbol{a}_1), \ldots, \psi_k(x_k, \boldsymbol{a}_k))$
iff $(A, \psi_1[\mathcal{A}, \boldsymbol{a}_1], \ldots, \psi_k[\mathcal{A}, \boldsymbol{a}_k]) \in \mathcal{K}$,
where $\psi_i[\mathcal{A}, \boldsymbol{a}_i] = \{a \in A \mid \mathcal{A} \models \psi_i(a, \boldsymbol{a}_i)\}$.

Here the tuple $\boldsymbol{a}_i$ gives the interpretation for those free variables in $\psi_i(x_i, \boldsymbol{y}_i)$ which are not equal to x_i. The extension $\text{FO}(\mathbf{Q})$ by a set $\mathbf{Q}$ of unary quantifiers is defined similarly by adding to FO the above rule for each $Q_\mathcal{K} \in \mathbf{Q}$. The quantifier rank of a formula of $\text{FO}(\mathbf{Q})$ is defined as usually by the maximum depth of nesting of quantifiers (counting both first-order and generalized quantifiers). We also write $\text{FO}(\mathbf{Q}_u)$ for FO extended by *all* (continuum many) unary quantifiers. Note that the same definition can be used with other ambient logics, not just FO. In particular, we shall use $\mathcal{L}_{\infty\omega}(\mathbf{Q}_u)$, the infinitary logic extended with unary generalized quantifiers.

We now list some well-known examples.

Example 1. (1) The existential quantifier $\exists$ corresponds to the class of structures $\{(A, P) \mid \emptyset \neq P \subseteq A\}$. Similarly, the universal quantifier $\forall$ can be identified with the unary quantifier which is defined by the class $\{(A, P) \mid P = A\}$.
(2) *Counting quantifier* $\exists^{\geq k}$ can be defined by the class $\{(A, P) \mid card(P) \geq k\}$. Note that each $\exists^{\geq k}$ can be defined in first-order logic; however, this requires k quantifiers $\exists$, and increases the quantifier rank by k. In contrast, the counting quantifier $\exists^{\geq k}$ increases the quantifier rank by one. We denote the set of all counting quantifiers $\exists^{\geq k}$ by $\mathbf{C}$, i.e., $\mathbf{C} = \{\exists^{\geq k} \mid k \geq 1\}$.
(3) *Counting modulo m quantifier* D_m is defined by the class $\{(A, P) \mid card(P) \equiv 0 \pmod{m}\}$. An easy first-order Ehrenfeucht-Fraïssé game argument shows that D_m is not definable in FO whenever $m > 1$. Thus $\text{FO}(D_m)$ is strictly more expressive than FO. For instance, the sentence $D_2 x(x = x)$ of $\text{FO}(D_2)$ says that the number of points in a structure is even; it is well-known that this property is not definable in first-order logic alone.
(4) *Majority quantifier* MAJ, which is defined by the class $\{(A, P) \mid card(P) \geq \frac{1}{2} card(A)\}$, is not definable in FO, either. For example, in $\text{FO}(\text{MAJ})$ we can say that there is a node in a graph $\mathcal{A}$ that is connected to at least half of the nodes of $\mathcal{A}$: the defining sentence is $\exists y \text{MAJ} x E(y, x)$. This quantifier is also interesting in connection with capturing complexity classes, see Section 6.
(5) Extending FO with *Rescher* (bigger cardinality) or *Härtig* (equicardinality) quantifiers also increases the expressive power. Rescher quantifier R is defined by the class $\{(A, P, S) \mid card(P) \leq card(S)\}$ and Härtig quantifier H by the class $\{(A, P, S) \mid card(P) = card(S)\}$. Thus, for instance, given two points a and b in an undirected graph, $Rx, y(E(a, x), E(b, y))$ says that a has at most as many neighbors as b. Similarly, $Hx, y(E(a, x), E(b, y))$ says that a and b have equally many neighbors.

For each $Q_{\mathcal{K}}$, there is a natural Ehrenfeucht-Fraïssé style game-theoretic characterization for elementary equivalence in $\mathrm{FO}(Q_{\mathcal{K}})$. We formulate the rules of this game for the counting modulo m quantifier. The rules of the game are as for first-order Ehrenfeucht-Fraïssé game except that now the spoiler may also choose a subset from one of the structures, say $X \subseteq A$. The duplicator has to respond by choosing a subset of the other structure, $Y \subseteq B$, which has modulo m equal cardinality to the spoiler's choice, that is, $card(Y) \equiv card(X) \pmod{m}$. The spoiler then challenges the duplicator's choice by selecting a point from the duplicator's structure, $b \in B$, and the duplicator has to choose a point from the other structure, $a \in A$, such that $a \in X$ if and only if $b \in Y$. Again, if $a_1, \dots, a_n$ are the points chosen from $\mathcal{A}$ and $b_1, \dots, b_n$ are the points chosen from $\mathcal{B}$ during n rounds, the duplicator wins if and only if $\{(a_i, b_i) \mid 1 \leq i \leq n\}$ is a partial isomorphism $\mathcal{A} \to \mathcal{B}$. We call this game the *counting modulo m Ehrenfeucht-Fraïssé* game.

Note that the first-order Ehrenfeucht-Fraïssé game can be seen as a special case of the counting modulo m Ehrenfeucht-Fraïssé game: the spoiler can choose the empty subset from one of the structures, and, in order to win, the duplicator has to respond with the empty set. The spoiler and the duplicator then choose their points from the complement of the empty set, that is, without any restrictions, just as in the regular Ehrenfeucht-Fraïssé game.

The following theorem shows that the counting modulo m Ehrenfeucht-Fraïssé game indeed gives us the game-theoretic characterization we were looking for. We use the notation $\mathcal{A} \equiv^n_{\mathrm{FO}(D_m)} \mathcal{B}$ when the duplicator has a winning strategy in the n-round counting modulo m Ehrenfeucht-Fraïssé game.

Theorem 2 (see [31]). *Let $\mathcal{A}, \mathcal{B} \in \mathrm{STRUCT}[\sigma]$. Then $\mathcal{A} \equiv^n_{\mathrm{FO}(D_m)} \mathcal{B}$ if and only if $\mathcal{A}$ and $\mathcal{B}$ agree on all $\mathrm{FO}(D_m)$ sentences of quantifier rank up to n.*   □

Corollary 2. *A class $\mathcal{C} \subseteq \mathrm{STRUCT}[\sigma]$ is not definable in $\mathrm{FO}(D_m)$ if and only if for every n there are σ-structures $\mathcal{A} \in \mathcal{C}$ and $\mathcal{B} \notin \mathcal{C}$ such that the duplicator has a winning strategy in the n-round counting modulo m Ehrenfeucht-Fraïssé game on $\mathcal{A}$ and $\mathcal{B}$.*

The intuition behind this theorem is that if all subsets of $\mathcal{A}$ and $\mathcal{B}$ that the spoiler and the duplicator can use in the game look similar, and there are modulo m equally many of them, then $\mathrm{FO}(D_m)$ can distinguish between $\mathcal{A}$ and $\mathcal{B}$ no more than FO can. In the next section we give a precise formulation for this intuition.

Theorem 2 can be used to show limits of expressive bounds of counting modulo quantifiers. For instance, we can show that connectivity of finite graphs is not definable in $\mathrm{FO}(D_m)$, for any m. The construction is similar to the first-order case, but we also have to require that there are modulo m equally many points in both structures. Then we can proceed very much like in the proof for first-order. Similarly, we can show that the majority quantifier MAJ (or Rescher or Härtig quantifiers) cannot be defined in $\mathrm{FO}(D_m)$. To see this, we can, for instance, take, for each n, $\mathcal{A} = (A, U)$ to be a set with a unary relation U

such that $card(A) = 3nm$ and $card(U) = 2nm$. Take $\mathcal{B} = (B, U)$ which satisfies $card(B) = 3nm$ but $card(U) = nm$. It is not difficult to show that the duplicator has a winning strategy in the n-round counting modulo m Ehrenfeucht-Fraïssé game over $\mathcal{A}$ and $\mathcal{B}$. But obviously $\mathcal{A} \models \mathrm{MAJ}xU(x)$ and $\mathcal{B} \not\models \mathrm{MAJ}xU(x)$. These results can be extended to the ordered case, see [41].

If we want to give a game-theoretic method to prove expressive bounds for first-order logic with *all* unary quantifiers, different techniques must be used. The method we employ here is based on *bijective Ehrenfeucht-Fraïssé* games. The rules of the game are the following. As before, the players are the spoiler and the duplicator. There are two σ-structures $\mathcal{A}$ and $\mathcal{B}$ and the number of rounds, say n, given. In each round i the duplicator first selects a bijection $f_i : A \to B$ (if $card(A) \neq card(B)$, then the duplicator loses), and then the spoiler selects a point $a_i \in A$. The duplicator has to select the point $f_i(a_i)$ from $\mathcal{B}$. This continues for n rounds. After the last round, the duplicator is declared the winner if and only if $\{(a_i, f_i(a_i)) \mid 1 \leq i \leq n\}$ is a partial isomorphism from $\mathcal{A}$ to $\mathcal{B}$. We use the notation $\mathcal{A} \equiv_{bij}^n \mathcal{B}$ if the duplicator has a winning strategy in the n-move bijective game on $\mathcal{A}$ and $\mathcal{B}$.

It turns out that this game characterizes elementary equivalence in a logic that is stronger than $\mathrm{FO}(\mathbf{Q}_u)$. This logic is obtained from $\mathrm{FO}(\mathbf{Q}_u)$ by allowing infinite disjunctions and conjunctions, but by keeping quantifier rank bounded. More precisely, let $\mathcal{L}_{\infty\omega}(\mathbf{Q}_u)^k$ be the extension of $\mathrm{FO}(\mathbf{Q}_u)$ where infinite disjunctions and conjunctions are allowed but quantifier rank of each formula is at most k. The union of all these logics $\mathcal{L}_{\infty\omega}(\mathbf{Q}_u)^k$ over all natural numbers $k < \omega$ is denoted by $\mathcal{L}_{\infty\omega}(\mathbf{Q}_u)^\omega$ (that is, the depth of nesting of quantifiers in each formula is finite). Methods used in [23] give us the following result (a proof can be found in [24]).

Theorem 3 (see [23,24]). *Let $\mathcal{A}, \mathcal{B} \in \mathrm{STRUCT}[\sigma]$. Then $\mathcal{A} \equiv_{bij}^n \mathcal{B}$ if and only if $\mathcal{A}$ and $\mathcal{B}$ agree on all $\mathcal{L}_{\infty\omega}(\mathbf{Q}_u)^\omega$ sentences of quantifier rank up to n.* □

Corollary 3. *A class $\mathcal{C} \subseteq \mathrm{STRUCT}[\sigma]$ is not definable in $\mathcal{L}_{\infty\omega}(\mathbf{Q}_u)^\omega$ if and only if for every n there are $\mathcal{A} \in \mathcal{C}$ and $\mathcal{B} \notin \mathcal{C}$ such that the duplicator has a winning strategy in the n-round bijective Ehrenfeucht-Fraïssé game on $\mathcal{A}$ and $\mathcal{B}$.* □

Note that the expressive power of $\mathrm{FO}(\mathbf{Q}_u)$ is strictly weaker than that of $\mathcal{L}_{\infty\omega}(\mathbf{Q}_u)^\omega$. In $\mathrm{FO}(\mathbf{Q}_u)$, it is not possible to express the second vectorization of Härtig quantifier (that is, the equicardinality quantifier for pairs) [39], while techniques used in [31] show that every vectorization of every unary quantifier can be defined in $\mathcal{L}_{\infty\omega}(\mathbf{Q}_u)^\omega$. It also follows from [31] that $\mathcal{L}_{\infty\omega}(\mathbf{Q}_u)^\omega$ is as strong in expressive power as $\mathcal{L}_{\infty\omega}(\mathbf{C})^\omega$. (In [31], this was shown for finite-variable logics, but the same proof technique works for $\mathcal{L}_{\infty\omega}(\mathbf{Q}_u)^\omega$ [23].)

Although $\mathcal{L}_{\infty\omega}(\mathbf{Q}_u)^\omega$ has strong counting power, the game characterization can be used to show that connectivity and transitive closure are not definable in it. The idea is the same as before: For each n, we can take $\mathcal{A}$ to be a chordless cycle of length 2^{n+1} and $\mathcal{B}$ a disjoint union of two chordless cycles of length 2^n.

Now the duplicator's strategy is to choose in round j a bijection that preserves the distances up to 2^{n-j} between the points next to the already chosen ones. A combinatorial proof can be given that shows the existence of such a strategy [27]; however, we shall see a much easier way to establish this in the next section.

We remark that bijective games are useless in the presence of a linear order. In order to win, the duplicator has to follow the linear order when he chooses bijections (otherwise the spoiler wins in the next two rounds). Thus, there is essentially only one bijection the duplicator can choose. This gives us an example of difficulties that arise when one attempts to prove expressivity bounds in the ordered setting. In the subsequent sections, we shall face similar problems several times.

Another interesting counting logic is FO + COUNT, cf. [15,30]. This is a two-sorted logic with the second sort being the sort of natural numbers. More precisely, in this approach a structure is of the form

$$\mathcal{A} = (\{v_1, \dots, v_n\}, \{1, \dots, n\}, \overline{R}_1, \dots, \overline{R}_l, \leq, BIT, 1, \max).$$

Here relations R_i apply to the non-numerical domain $\{v_1, \dots, v_n\}$, while the linear order $\leq$, the BIT predicate and the constants 1 and max (interpreted as 1 and n) refer to the numerical domain $\{1, \dots, n\}$. Here $BIT(i,j)$ holds if and only if the ith bit in the binary representation of j is one. These two disjoint domains are connected by allowing formulas of the form $\exists ix\varphi(x)$ with the semantics that at least i elements satisfy φ, i.e., $card(\{a \in A \mid \mathcal{A} \models \varphi(a)\}) \geq i$. Here i refers to the numerical domain and x refers to the non-numerical domain; the quantifier $\exists ix$ binds x but not i. As an example, consider the sentence

$$\exists i \exists j[(j+j=i) \land \exists ix\varphi(x) \land \forall k(\exists kx\varphi(x) \to k \leq i)].$$

This sentence tests if the cardinality of $\{a \mid \varphi(a)\}$ is even. Indeed, $\exists ix\varphi(x) \land \forall k(\exists kx\varphi(x) \to k \leq i)$ holds iff exactly i elements satisfy φ, and i is even since it is of the form $2j$. Note that we used the fact that $+$ and $*$ are definable as ternary predicates in the presence of the BIT relation, cf. [15,28].

Remark 1. L. Hella made the following observation. While first-order logic extended by Rescher quantifier R is in general strictly weaker than FO + COUNT, in the presence of a built-out linear order, FO(R) and FO + COUNT have the same expressive power. Clearly, $Rx, y(\varphi(x), \psi(y))$ can be written in FO+COUNT. On the other hand, $\exists ix\varphi(x)$ can be expressed by $Rj, x(j < i, \varphi(x))$.

A game-theoretic characterization for elementary equivalence in FO+COUNT was introduced in [30] and used subsequently in [14]. However, we do not go into detail here, mainly due to the fact that the counting games of [30] are subsumed by the bijective games. The logic FO + COUNT has a number of applications in computer science, in particular, in complexity theory. This will be discussed in Section 6.

Finally, we refer the reader to [31,48] for a more detailed overview of results on generalized quantifiers in finite-model theory.

4   Gaifman's and Hanf's Conditions

The game-theoretic characterizations for elementary equivalence of logics considered in the previous section gave us a vague intuition that these logics can only express local properties. This intuition will be formalized in this section. We review theorems by Gaifman and Hanf, and their modifications.

4.1   Gaifman's Theorem

We start with Gaifman's theorem [18]. Let $\mathcal{A}$ be a σ-structure. Recall that the distance $d(a, b)$ is the length of the shortest path from a to b in the Gaifman graph of $\mathcal{A}$, and $S_r(\boldsymbol{a}) = \{b \mid d(b, a) \leq r, a \in \boldsymbol{a}\}$. For each fixed k, there are first-order formulas that define the relations $d(a, b) > k$, $d(a, b) = k$ and $d(a, b) < k$ (see [13]). Hence, bounded quantifications of the form $\forall x \in S_k(\boldsymbol{y})$ and $\exists x \in S_k(\boldsymbol{y})$ are expressible in first-order logic for every fixed k. A formula $\varphi^{(r)}(\boldsymbol{y})$ is called *r-local around* $\boldsymbol{y}$ if every quantifier in it is of the form $\forall x \in S_k(\boldsymbol{y})$ or $\exists x \in S_k(\boldsymbol{y})$ with $k \leq r$. A sentence ψ is called *basic r-local* if it is of the form

$$\exists x_1 \ldots \exists x_m \left(\bigwedge_{1 \leq i \leq m} \varphi^{(r)}(x_i) \wedge \bigwedge_{1 \leq i < j \leq m} d(x_i, x_j) > 2r \right)$$

where $\varphi^{(r)}(x)$ is an r-local formula around x.

Theorem 4 (Gaifman [18]). *Every first-order sentence is equivalent to a Boolean combination of basic r-local sentences, and every first-order formula $\varphi(x_1, \ldots$ $\ldots, x_n)$ is equivalent to a Boolean combination of t-local formulas around $x_1, \ldots$ $\ldots, x_n$ and basic r-local sentences. Furthermore, $r \leq 7^{qr(\varphi)-1}$, $t \leq (7^{qr(\varphi)-1})/2$ and $m \leq n + qr(\varphi)$.* $\square$

Gaifman's theorem tells us that every first-order formula can see only a bounded number of local neighborhoods in a structure, i.e., only a small part of the input. This is indeed a formalization of the informal statement that first-order logic can express only local properties.

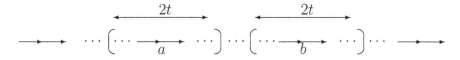

Fig. 1. Formula $\varphi(x, y)$ cannot distinguish (a, b) from (b, a).

Example 2. We show that first-order logic cannot express the transitive closure of a directed graph. Assume, to the contrary, that there is a first-order formula $\varphi(x, y)$ in the language of a single binary relation E, such that $\varphi[\mathcal{A}] =$

$(A, \{(a, b) \in A \mid A \models \varphi(a, b)\})$ is the transitive closure of $\mathcal{A}$. Apply Gaifman's theorem to it, and find t and r. Now consider the graph shown in Figure 1. It is a successor relation, on which we select two points, a and b. Assume that $d(a, b) > 2t$, and the distances from a and b to the start and the end node of the graph are at least $t + 1$. Then the t-neighborhoods of (a, b) and (b, a) are isomorphic, and by Gaifman's theorem, φ cannot distinguish (a, b) from (b, a). Thus, φ cannot define the transitive closure, since $(a, b) \in TRCL(\mathcal{A})$, but $(b, a) \notin TRCL(\mathcal{A})$.

4.2 Hanf's Theorem and its Modifications

While Gaifman's theorem helps prove expressivity bounds for FO directly, without resorting to establishing a winning strategy for the duplicator[1], Hanf's theorem [22] and its numerous modifications [17,24,40,41] provide criteria for the existence of a strategy for the duplicator that is based on counting of small neighborhoods in two structures.

Hanf's theorem was originally proved for infinite structures. It was observed by Fagin, Stockmeyer and Vardi [17] that the technique can be modified to be applicable to finite structures. The extensions of Hanf's technique [24,40,41] follow the ideas of [17].

Let $\mathcal{A}$ be a σ-structure and $a \in A$. Recall that the isomorphism type of $N_d^{\mathcal{A}}(a)$ is denoted by $tp_d^{\mathcal{A}}(a)$. Let τ be an isomorphism type of a σ_1-structure (σ extended with one constant). We denote the number of points $a \in A$ whose d-neighborhoods realize τ by $n_d(\mathcal{A}, \tau)$. That is,

$$n_d(\mathcal{A}, \tau) \;=\; card(\{a \in A \mid tp_d^{\mathcal{A}}(a) = \tau\}).$$

For example, if $\mathcal{A}$ is a chordless undirected cycle of length at least $2d + 2$ then there is only one isomorphism type τ of a d-neighborhood of a point occurring in $\mathcal{A}$: the chordless path of length $2d + 1$. In this case $n_d(\mathcal{A}, \tau) = card(A)$.

We call structures $\mathcal{A}$ and $\mathcal{B}$ (d, m)-*equivalent* if for every isomorphism type τ they have exactly the same number of points whose d-neighborhoods realize τ, or both structures have at least m such points, that is,

$$\min(n_d(\mathcal{A}, \tau), m) \;=\; \min(n_d(\mathcal{B}, \tau), m).$$

The modification of Hanf's theorem for the finite case is the following.

Theorem 5 ([17]). *Let n and f be positive integers. There are positive integers d and m such that whenever $\mathcal{A}$ and $\mathcal{B}$ are (d, m)-equivalent structures where every point has degree at most f, then $\mathcal{A} \equiv_{FO}^n \mathcal{B}$, that is, $\mathcal{A}$ and $\mathcal{B}$ satisfy the same sentences of FO of quantifier rank up to n.* □

---

[1] New winning conditions for the duplicator based on Gaifman's theorem were presented recently in [47].

Note that since we consider finite models, for any $\mathcal{A}$ there is a number f that exceeds degrees of all points of $\mathcal{A}$. This leads to the following. We say that $\mathcal{A}$ and $\mathcal{B}$ are *d-equivalent*, written as $\mathcal{A} \leftrightarrows_d \mathcal{B}$, if for every type of a d-neighborhood of a point, τ, equally many points realize it in $\mathcal{A}$ and $\mathcal{B}$. That is,

$$\mathcal{A} \leftrightarrows_d \mathcal{B} \quad \text{iff} \quad n_d(\mathcal{A}, \tau) = n_d(\mathcal{B}, \tau) \text{ for every } \tau.$$

Corollary 4 ([17]). *Let n be a positive integer. There there exists a positive integer d such that $\mathcal{A} \leftrightarrows_d \mathcal{B}$ implies $\mathcal{A} \equiv^n_{FO} \mathcal{B}$.* $\square$

This result makes precise the intuition that counting power of first-order logic is rather limited. It also shows that only local neighborhoods are relevant for elementary equivalence in FO. Most importantly, the result above yields much simpler proofs of expressivity bounds that those based on games. Below we give a canonical example of applicability of Hanf's technique.

Example 3. We show that connectivity of finite graphs is not definable in FO. Assume, to the contrary, that it is definable by a FO sentence Φ of quantifier rank n. Apply Corollary 4 to find $d > 0$ such that $\mathcal{A} \leftrightarrows_d \mathcal{B}$ would imply $\mathcal{A} \models \Phi$ iff $\mathcal{B} \models \Phi$. Now let $\mathcal{A}$ be a (chordless) cycle which has length $4d + 4$, and let $\mathcal{B}$ be a disjoint union of two chordless cycles of length $2d + 2$, see Figure 2.

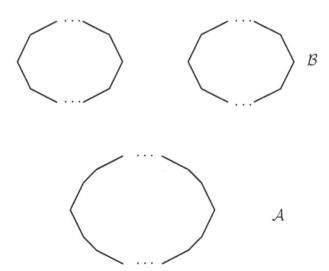

Fig. 2. Hanf's technique proves that connectivity is not first-order

As we noticed before, there is only one type of a d-neighborhood that these graphs realize, namely a chain on $2d + 1$ points. Thus, $\mathcal{A} \leftrightarrows_d \mathcal{B}$, since they have the same number of vertices. At the same time, $\mathcal{A}$ is connected, but $\mathcal{B}$ is not, proving that connectivity cannot be expressed by Φ.

Other examples, proved previously with games, can be shown to be derivable from Hanf's technique. For example, with the concept of (d, m)-equivalence, it is now easy to show that the majority quantifier (or Rescher and Härtig quantifiers) are not definable in first-order logic.   □

Before we outline the proof of Theorem 5 in [17], we state Hanf's theorem from [22]. It says that two arbitrary structures $\mathcal{A}$ and $\mathcal{B}$ are elementary equivalent if $S_r^{\mathcal{A}}(a)$ and $S_r^{\mathcal{B}}(b)$ are finite, for every finite r and every $a \in A$ and $b \in B$, and, for each $r > 0$ and each type τ of an r-neighborhood of a point, either $n_r(\mathcal{A}, \tau) = n_r(\mathcal{B}, \tau) < \omega$, or both $\mathcal{A}$ and $\mathcal{B}$ have infinitely many realizers of type τ.

Now we explain how Theorem 5 is proved. First, d is taken to be $\frac{3^n-1}{2}$, and m is taken to exceed the size of any d-neighborhood of a point in a structure whose degrees are bounded by f. Define $d_j = \frac{3^{n+1-j}-1}{2}$ for $j \leq n$; in particular, $d_j = 3 \cdot d_{j+1} + 1$, $d_1 = d$ and $d_n = 1$. It is then shown that the duplicator can play in the first-order Ehrenfeucht-Fraïssé game in such a way that after each round $j \leq n$, if $\boldsymbol{a} \in A^j$ and $\boldsymbol{b} \in B^j$ are the points chosen so far, then $tp_{d_j}^{\mathcal{A}}(\boldsymbol{a}) = tp_{d_j}^{\mathcal{B}}(\boldsymbol{b})$. This suffices, since $tp_1^{\mathcal{A}}(a_1, \dots, a_n) = tp_1^{\mathcal{B}}(b_1, \dots, b_n)$ implies that the mapping $a_i \mapsto b_i$ is a partial isomorphism. The condition that $\mathcal{A}$ and $\mathcal{B}$ are (d, m)-equivalent allows the duplicator to make the first move by choosing a point with a given type of its d-neighborhood. Suppose that the duplicator maintained the condition successfully for the first j rounds. That is, we have $tp_{d_j}^{\mathcal{A}}(\boldsymbol{a}) = tp_{d_j}^{\mathcal{B}}(\boldsymbol{b})$. Assume that $a \in A$ is spoiler's next choice in the game. If $a \in S_{2d_{j+1}+1}^{\mathcal{A}}(\boldsymbol{a})$, then the duplicator selects the point $b \in S_{2d_{j+1}+1}^{\mathcal{B}}$ which is given by the isomorphism between $N_{d_j}^{\mathcal{A}}(\boldsymbol{a})$ and $N_{d_j}^{\mathcal{B}}(\boldsymbol{b})$. Then one easily checks that $tp_{d_{j+1}}^{\mathcal{A}}(\boldsymbol{a}a) = tp_{d_{j+1}}^{\mathcal{B}}(\boldsymbol{b}b)$. If $a \notin S_{2d_{j+1}+1}^{\mathcal{A}}(\boldsymbol{a})$ then the duplicator can choose any $b \notin S_{2d_{j+1}+1}^{\mathcal{B}}(\boldsymbol{b})$ such that $tp_{d_{j+1}}^{\mathcal{A}}(a) = tp_{d_{j+1}}^{\mathcal{B}}(b)$. If m is chosen as above, such a point b always exists. Then again we can see that $tp_{d_{j+1}}^{\mathcal{A}}(\boldsymbol{a}a) = tp_{d_{j+1}}^{\mathcal{B}}(\boldsymbol{b}b)$ holds, since no points in $S_{d_{j+1}}^{\mathcal{A}}(\boldsymbol{a})$ and $S_{d_{j+1}}^{\mathcal{A}}(a)$ belong to the same tuple of an $\mathcal{A}$-relation, and likewise for $\mathcal{B}$.

Having established Hanf's condition for FO, we turn to counting extensions of FO. In [41], the proof of Theorem 5 was modified for $\mathrm{FO}(D_k)$, $k > 1$. Two structures $\mathcal{A}$ and $\mathcal{B}$ are called (d, m, D_k)-equivalent, if for each isomorphism type τ they have the same number of points whose d-neighborhoods realize τ, or in both structures there are at least m such points but modulo k equally many, that is,

$$\min(n_d(\mathcal{A}, \tau), m) = \min(n_d(\mathcal{B}, \tau), m) \quad \text{and} \quad n_d(\mathcal{A}, \tau) \equiv n_d(\mathcal{B}, \tau) \pmod{k}.$$

Note that (d, m, D_1)-equivalence is just the (d, m)-equivalence. Thus, the following result extends Theorem 5.

Theorem 6 ([41]). *Let n, f and k be positive integers. There are positive integers d and m such that whenever $\mathcal{A}$ and $\mathcal{B}$ are (d, m, D_k)-equivalent structures where every point has degree at most f, then $\mathcal{A}\equiv_{\mathrm{FO}(D_k)}^n \mathcal{B}$. That is, $\mathcal{A}$ and $\mathcal{B}$ satisfy the same sentences of $\mathrm{FO}(D_k)$ of quantifier rank up to n.*   □

The intuition behind this theorem is that the counting power of $FO(D_k)$ is rather limited, and that $FO(D_k)$ is not much stronger than FO. For instance, we can use Theorem 6 to show that connectivity is not definable in $FO(D_k)$ for any k, by using the same argument as in Example 3.

The proof of Theorem 6 in [41] follows the same idea than the proof of Theorem 5. One could ask whether this proof technique can be easily modified to prove elementary equivalence for $FO(Q)$, where Q is an arbitrary unary quantifier. It turns out that in this technique additivity property of the quantifier is crucial, and all unary quantifiers which satisfy this property are essentially counting modulo quantifiers. However, we now show a more general result that describes a winning condition in a *bijective* game.

Theorem 7 ([24,40]). *Let n be a positive integer. There is a positive integer d such that $\mathcal{A} \leftrightarrows_d \mathcal{B}$ implies $\mathcal{A} \equiv^n_{bij} \mathcal{B}$. In particular, if $\mathcal{A} \leftrightarrows_d \mathcal{B}$, then $\mathcal{A}$ and $\mathcal{B}$ satisfy the same sentences of $\mathcal{L}_{\infty\omega}(\mathbf{Q}_u)$ of quantifier rank up to n.* □

Before we explain how this result is proved, we give the following alternative definition of d-equivalence. Two structures $\mathcal{A}$ and $\mathcal{B}$ are d-equivalent if there exists a bijection $f : A \rightarrow B$ such that for every $a \in A$,

$$tp^{\mathcal{A}}_d(a) = tp^{\mathcal{B}}_d(f(a)).$$

Now the proof of Theorem 7 is very similar to the proof of Hanf's theorem for first-order logic. We again let d_j be $\frac{3^{n+1-j}-1}{2}$, and take d to be d_1. The duplicator's strategy is to play so that after each round j in the bijective Ehrenfeucht-Fraïssé game we have $tp^{\mathcal{A}}_{d_j}(\boldsymbol{a}) = tp^{\mathcal{B}}_{d_j}(\boldsymbol{b})$, if $\boldsymbol{a} \in A^j$ and $\boldsymbol{b} \in B^j$ are chosen during those j rounds. The first round bijection is given by the reformulation of d-equivalence above. Assume j rounds have been played, and we have $tp^{\mathcal{A}}_{d_j}(\boldsymbol{a}) = tp^{\mathcal{B}}_{d_j}(\boldsymbol{b})$, where $d_j = 3 \cdot d_{j+1} + 1$. Assume that $tp^{\mathcal{A}}_{d_{j+1}}(a) = \tau$, for $a \in S^{\mathcal{A}}_{2d_{j+1}+1}(\boldsymbol{a})$. Then $tp^{\mathcal{B}}_{d_{j+1}}(h(a)) = \tau$, and $h(a) \in S^{\mathcal{B}}_{2d_{j+1}+1}(\boldsymbol{b})$, where $h : N^{\mathcal{A}}_{d_j}(\boldsymbol{a}) \rightarrow N^{\mathcal{B}}_{d_j}(\boldsymbol{b})$ is an isomorphism. Thus, the number of points realizing τ is the same in complements of $S^{\mathcal{A}}_{2d_{j+1}+1}(\boldsymbol{a})$ and $S^{\mathcal{B}}_{2d_{j+1}+1}(\boldsymbol{b})$, and hence we have a bijection $g : A - S^{\mathcal{A}}_{2d_{j+1}+1}(\boldsymbol{a}) \rightarrow B - S^{\mathcal{B}}_{2d_{j+1}+1}(\boldsymbol{b})$ with the property $tp^{\mathcal{A}}_{d_{j+1}}(x) = tp^{\mathcal{B}}_{d_{j+1}}(g(x))$. We now define a bijection f_{j+1} for the round $j+1$ to be h on $S^{\mathcal{A}}_{2d_{j+1}+1}(\boldsymbol{a})$, and g on its complement. It is routine to verify that f_{j+1} is a bijection, and that $tp^{\mathcal{A}}_{d_{j+1}}(\boldsymbol{a}a) = tp^{\mathcal{B}}_{d_{j+1}}(\boldsymbol{b}f_{j+1}(a))$ for every $a \in A$. After the last round we have vectors with isomorphic 1-neighborhoods; hence they define a partial isomorphism.

Theorem 7 shows the limits of expressive power provided by all unary quantifiers. It also significantly simplifies proofs of expressivity bounds, as applying bijective games is usually not a very easy task. For instance, we can use Example 3 to show that connectivity of finite graphs is not definable in $\mathcal{L}_{\infty\omega}(\mathbf{Q}_u)^\omega$, thus avoiding all the tedious combinatorics involved in an argument based on bijective games.

As another example, we consider classes of undirected graphs which are *closed under stretching*. More precisely, let $\mathcal{A}$ be a graph and let $\mathcal{A}_d$ be the graph obtained by replacing every edge of $\mathcal{A}$ by a path of length $2d + 1$. Then each d-neighborhood of a point in $\mathcal{A}_d$ contains at most one node whose degree is greater than two. We say that a class $\mathcal{C}$ of graphs is closed under stretching if $\mathcal{A}_d \in \mathcal{C}$ for every $\mathcal{A} \in \mathcal{C}$ and for every positive integer d. Now it is easy to see that if $\mathcal{A}$ and $\mathcal{B}$ are 1-equivalent graphs, then $\mathcal{A}_d$ and $\mathcal{B}_d$ are d-equivalent. In other words, if there are the same number of points in both graphs $\mathcal{A}$ and $\mathcal{B}$ of each degree, then $\mathcal{A}_d$ and $\mathcal{B}_d$ are d-equivalent. It can then be proved that if a class $\mathcal{C}$ and its complement are closed under stretching, and there are $\mathcal{A} \in \mathcal{C}$ and $\mathcal{B} \notin \mathcal{C}$ as above, then $\mathcal{C}$ is not definable in $\mathcal{L}_{\infty\omega}(\mathbf{Q}_u)^\omega$ [26]. This argument shows that graph properties such as planarity and 3-colorability are not definable in $\mathcal{L}_{\infty\omega}(\mathbf{Q}_u)^\omega$.

Hanf's technique was also used by Etessami [15] (although a preliminary conference version [14] had a proof based on counting games of [30]). It was shown in [15] that a linear order cannot be defined in FO + COUNT from its underlying successor relation. The proof relies on Hanf's technique for FO + COUNT. Its applicability follows from Theorem 7, since every FO + COUNT sentence is definable in $\mathcal{L}_{\infty\omega}(\mathbf{Q}_u)^\omega$.

Summing up, the combinatorial arguments in this section allow us to simplify many proofs that were originally shown by using difficult game-theoretic arguments. Furthermore, we can also prove, often quite easily, new nondefinability results. Note, however, that in the presence of a linear ordering, none of these results can be applied to derive inexpressibility results. This is because in the presence of an order, every point is a neighbor of any other point, and thus $S_1^{\mathcal{A}}(a)$ contains all elements of $\mathcal{A}$. Many results proving expressive bounds of these logics on ordered structures still use games (see [13] for first-order case and [41] for FO(D_m)). However, bijective Ehrenfeucht-Fraïssé games cannot be applied at all in the ordered case. In fact, this is closely connected to some deep problems in circuit complexity, as will be explained in Section 6.

5   Abstract Notions of Locality

On the surface, Gaifman's theorem and various forms of Hanf's theorem appear to be quite unrelated. Nevertheless, we shall see soon that there is a very close relationship between these results. To make statements like this, we study abstract concepts behind locality theorems. This approach can be seen as the study of the essential ideas behind the proofs of locality theorems, rather than using the statements of the theorems for proving expressivity bounds. From these theorems, we extract abstract notions of locality, and show how they are related. We also discuss a new form of locality, the bounded degree property, and show how it is related to other forms. This property turns out to be particularly simple to use in proving expressivity bounds.

5.1 Gaifman's Locality

We start by analyzing Gaifman's theorem. This theorem says that only local neighborhoods are important for elementary equivalence in first-order logic. This is captured by the following definition.

Definition 1 ([32,24]). *A formula* $\psi(x_1, \ldots, x_m)$ *is* Gaifman-local *if there exists* $r > 0$ *such that for every* $\mathcal{A} \in \mathrm{STRUCT}[\sigma]$ *and for every two* m-*ary vectors* $a, b \in A^m$,

$$tp_r^{\mathcal{A}}(a) = tp_r^{\mathcal{A}}(b) \quad \text{implies} \quad \mathcal{A} \models \psi(a) \text{ if and only if } \mathcal{A} \models \psi(b).$$

The minimum r *for which this holds is called the* locality rank *of* ψ *and is denoted by* $\mathrm{lr}(\psi)$.

This definition formulates that if a logic is Gaifman-local, i.e., every formula of a logic is Gaifman-local, then indeed only small parts of the input are relevant for elementary equivalence in the logic. The part of Gaifman's theorem that deals with sentences, gives rise to the following notion.

Definition 2 ([32,24]).

- *A formula* $\psi(x_1, \ldots, x_m)$, $m \geq 1$, *is* strongly Gaifman-local *if there exists* $r > 0$ *such that for every* $\mathcal{A}, \mathcal{B} \in \mathrm{STRUCT}[\sigma]$ *and for every two* m-*ary vectors* $a \in A^m$ *and* $b \in B^m$, $tp_r^{\mathcal{A}}(a) = tp_r^{\mathcal{B}}(b)$ *implies* $\mathcal{A} \models \psi(a)$ *if and only if* $\mathcal{B} \models \psi(b)$.
- *A sentence* Ψ *is* strongly Gaifman-local *if it is equivalent to a Boolean combination of sentences of the form* $\exists y \psi(y)$, *where* $\psi(y)$ *is a strongly Gaifman-local formula.*

Before going further we notice that not every first-order formula is *strongly* Gaifman local. Consider the class of directed graphs. Let $\psi(x)$ be the formula $\forall y \neg E(y, x) \wedge \exists z \forall y \neg E(z, y)$. Then $\psi(x)$ says that x has not a predecessor and there is a point which has no a successor. Suppose that ψ is strongly Gaifman-local with locality rank r. Let G_1 be a chain graph of length $r + 1$ and let G_2 be a chain graph of the same length with a loop on the end node. Denote the start node of G_i by a_i. Then $tp_r^{G_1}(a_1) = tp_r^{G_2}(a_2)$ but obviously $G_1 \models \psi(a_1)$ whereas $G_2 \not\models \psi(a_2)$.

From Gaifman's theorem, we can derive the following.

Corollary 5 ([18,32,24]). *Every first-order formula is Gaifman-local and every first-order sentence is strongly Gaifman-local. Moreover, for every* $\psi(x)$, $\mathrm{lr}(\psi) \leq (7^{qr(\psi)} - 1)/2$. $\square$

5.2 Hanf's Locality

When we studied Hanf's condition and proofs of Theorems 5, 6 and 7, we noticed that the essential part in all the proofs was the concept of d-equivalence. We now consider this concept in more detail and define the notion of Hanf's locality.

Before giving the definition of the abstract notion we need more notation and definitions.

We extend the notion of d-equivalence (see Section 4) for structures with parameters. Let $\mathcal{A}, \mathcal{B} \in \text{STRUCT}[\sigma]$ and $\boldsymbol{a} \in A^m$, $\boldsymbol{b} \in B^m$. Then $(\mathcal{A}, \boldsymbol{a})$ and $(\mathcal{B}, \boldsymbol{b})$ are d-equivalent, $(\mathcal{A}, \boldsymbol{a}) \leftrightarrows_d (\mathcal{B}, \boldsymbol{b})$, if for every isomorphism type τ,

$$n_d((\mathcal{A}, \boldsymbol{a}), \tau) = n_d((\mathcal{B}, \boldsymbol{b}), \tau).$$

In other words, there are as many points $a \in A$ and $b \in B$ such that $tp_d^{\mathcal{A}}(\boldsymbol{a}a) = tp_d^{\mathcal{B}}(\boldsymbol{b}b)$; or equivalently, there is a bijection $f : A \to B$ such that $tp_d^{\mathcal{A}}(\boldsymbol{a}x) = tp_d^{\mathcal{B}}(\boldsymbol{b}f(x))$ for all $x \in A$.

Definition 3 ([32,24]). *A formula* $\psi(x_1, \dots, x_m)$ *is* Hanf-local *if there exists a number* d *such that for every* $\mathcal{A}, \mathcal{B} \in \text{STRUCT}[\sigma]$ *and for every two m-ary vectors* $\boldsymbol{a}$ *and* $\boldsymbol{b}$ *of elements of* $\mathcal{A}$ *and* $\mathcal{B}$ *respectively,*

$$(\mathcal{A}, \boldsymbol{a}) \leftrightarrows_d (\mathcal{B}, \boldsymbol{b}) \quad implies \quad \mathcal{A} \models \psi(\boldsymbol{a}) \text{ if and only if } \mathcal{B} \models \psi(\boldsymbol{b}).$$

The minimum d *for which this holds is called the* Hanf locality rank *of* ψ, *and is denoted by* $\text{hlr}(\psi)$.

Thus, a sentence φ is Hanf-local, if there exists a number d such that $\mathcal{A} \leftrightarrows_d \mathcal{B}$ implies $\mathcal{A} \models \varphi$ iff $\mathcal{B} \models \varphi$. From results of the previous section, the following theorem is immediate.

Theorem 8 ([32,24]). *Every sentence* φ *of* $\mathcal{L}_{\infty\omega}(\mathbf{Q}_u)^\omega$, *or* FO + COUNT, *or* FO *is Hanf-local, and* $\text{hlr}(\varphi) \leq 3^{qr(\varphi)}$.   □

This result confirms that although these logics have substantial counting power, they can only recognize properties of small neighborhoods, and cannot grasp a structure as a whole. The definitions above extract the essential notions that were crucial for proving theorems in the previous section. Now these notions allow us to *compare* different locality results.

5.3   Relationship between the Notions of Locality

The result below is the main technical lemma that establishes the relationship between strong Gaifman-locality, Gaifman-locality and Hanf-locality. It states that d-equivalence of structures extends to d-equivalence of structures with parameters, if large enough neighborhoods of the parameters are isomorphic.

Lemma 1 ([24]). *If* $\mathcal{A} \leftrightarrows_d \mathcal{B}$ *and* $tp_{3d+1}^{\mathcal{A}}(\boldsymbol{a}) = tp_{3d+1}^{\mathcal{B}}(\boldsymbol{b})$, *then* $(\mathcal{A}, \boldsymbol{a}) \leftrightarrows_d (\mathcal{B}, \boldsymbol{b})$.   □

Using this, we prove the following.

Theorem 9 ([32,24]). *Every Hanf-local formula is Gaifman-local.*

Proof. Suppose $\psi(x_1, \dots, x_m)$ is a Hanf-local formula with $\text{hlr}(\psi) = d$. We show that ψ is Gaifman-local. Take any two m-vectors $\boldsymbol{a}$ and $\boldsymbol{b}$ of a structure $\mathcal{A}$ such that $tp_r(\boldsymbol{a}) = tp_r(\boldsymbol{b})$, where $r = 3d + 1$. Since $\mathcal{A} \leftrightarrows_d \mathcal{A}$, by Lemma 1 we obtain $(\mathcal{A}, \boldsymbol{a}) \leftrightarrows_d (\mathcal{A}, \boldsymbol{b})$. Thus, $\mathcal{A} \models \psi(\boldsymbol{a})$ if and only if $\mathcal{A} \models \psi(\boldsymbol{b})$. Hence ψ is Gaifman-local and $\text{lr}(\psi) \leq 3d + 1$.   □

We now consider the relationship between Hanf's locality and strong Gaifman's locality. As a technical tool, we need to extend the notion of d-equivalence to tuples. The number of different m-tuples whose d-neighborhoods realize an isomorphism type τ^m of a σ_m-structure $\mathcal{A}$, is denoted by $n_d(\mathcal{A}, \tau^m)$. We write $\mathcal{A} \leftrightarrows_{m,d} \mathcal{B}$, if for every isomorphism type τ^m,

$$n_d(\mathcal{A}, \tau^m) = n_d(\mathcal{B}, \tau^m).$$

Equivalently, $\mathcal{A} \leftrightarrows_{m,d} \mathcal{B}$ if and only if there is a bijection $f : A^m \to B^m$ such that $tp_d^{\mathcal{A}}(a) = tp_d^{\mathcal{B}}(f(a))$ for every $a \in A^m$. By considering m-tuples whose components are the same, we see that for all $m > 0$, $\mathcal{A} \leftrightarrows_{m,d} \mathcal{B}$ implies $\mathcal{A} \leftrightarrows_d \mathcal{B}$.

Our crucial lemma is that r-equivalence of $(m+1)$-tuples can be guaranteed by d-equivalence of tuples for large enough d that depends only on r. This can be shown by considering parametrized versions of these structures and applying Lemma 1.

Proposition 1 ([32,24]). *Let $m > 0$ and $d \geq 0$. Then $\mathcal{A} \leftrightarrows_{m,3d+1} \mathcal{B}$ implies $\mathcal{A} \leftrightarrows_{m+1,d} \mathcal{B}$. In particular, for every $r > 0$ and $m \geq 1$ there is d such that $\mathcal{A} \leftrightarrows_d \mathcal{B}$ implies $\mathcal{A} \leftrightarrows_{m,r} \mathcal{B}$.* ☐

This can be used to show the following.

Theorem 10 ([32,24]). *Every strongly Gaifman-local sentence is Hanf-local.*
Consider a sentence Ψ which is equivalent to $\exists x_1 \ldots \exists x_m \psi(x_1, \ldots, x_m)$, where $\psi(x)$ is strongly Gaifman-local. Let $r > 0$ witness strong Gaifman's locality of ψ. Take d given by Proposition 1. Then $\mathsf{hlr}(\Psi) \leq d$. Indeed, let $\mathcal{A} \leftrightarrows_d \mathcal{B}$ and $\mathcal{A} \models \Psi$. Then $\mathcal{A} \models \psi(a)$ for some $a \in A^m$. By Proposition 1 we know that $\mathcal{A} \leftrightarrows_{m,r} \mathcal{B}$, and thus we find $b \in B^m$ such that $tp_r^{\mathcal{A}}(a) = tp_r^{\mathcal{B}}(b)$. Since ψ is strongly Gaifman-local, $\mathcal{B} \models \psi(b)$ and thus $\mathcal{B} \models \Psi$. Hence, $\mathsf{hlr}(\Psi) \leq d$. ☐

This implies that the two parts of Gaifman's theorem (those dealing with sentences and open formulas) are not independent. In fact, for any logic satisfying some regularity properties, strong Gaifman-locality of its sentences implies Gaifman-locality of its open formulae. See [24,32] for details.

5.4   Bounded Degree Property

One of the easiest ways to prove expressivity bounds is the *bounded degree property*. It was first introduced for graph queries in studying limits of expressive power of database query languages [36]. Later it was generalized to arbitrary (finite) structures in [11]. We now review this concept, show its usefulness in proving expressivity bounds, and relate it to other notions of locality.

For a relation $\overline{R}_i$ in $\mathcal{A}$, we define $\mathrm{degree}_j(R_i, a)$ to be the number of tuples in $\overline{R}_i$ whose jth component is a. For directed graphs, this gives us the familiar notions of in- and out-degree. The set

$$\{\mathrm{degree}_j(R_i, a) \mid R_i \in \sigma, a \in A, j \leq p_i\}$$

of all degrees realized in $\mathcal{A}$ is denoted by deg_set$(\mathcal{A})$. We use deg_count$(\mathcal{A})$ for $card($deg_set$(\mathcal{A}))$. The class of σ-structures $\mathcal{A}$ for which deg_set$(\mathcal{A}) \subseteq \{0, \ldots, k\}$ is denoted by STRUCT$_k[\sigma]$.

Informally, a query has the bounded degree property if an upper bound on the degrees in an input structure implies an upper bound on the number of degrees realized in the output structure produced by the query. Recall that the output of $\psi(x_1, \ldots, x_m)$ on $\mathcal{A}$, $\psi[\mathcal{A}]$, is the structure with one m-ary relation $(A, \{\boldsymbol{a} \in A^m \mid \mathcal{A} \models \psi(\boldsymbol{a})\})$.

Definition 4 ([11]). *A formula* $\psi(x_1, \ldots, x_m)$ *has the* bounded degree property *(BDP), if there is a function* $f_\psi : \mathbb{N} \to \mathbb{N}$ *such that*

$$deg\_count(\psi[\mathcal{A}]) \leq f_\psi(k) \text{ for any } \mathcal{A} \in \text{STRUCT}_k[\sigma].$$

The bounded degree property is a very useful tool in proving inexpressibility results of recursive properties, i.e., for those queries that require fixpoint computation. As a simple example, we show that the transitive closure query violates the BDP. Assume that *TRCL* does have the BDP; that is, there is a function $f : \mathbb{N} \to \mathbb{N}$ such that deg_count$(TRCL(\mathcal{A})) \leq f(k)$ if all in- and out-degrees in $\mathcal{A}$ do not exceed k. Let $N = f(1) + 1$. Consider $\mathcal{A}$ which is a successor relation on N points (see Figure 1). Since all in- and out-degrees in $\mathcal{A}$ are at most 1, we get deg_count$(TRCL(\mathcal{A})) \leq f(1) < N$, but one can easily see that deg_count$(TRCL(\mathcal{A})) = N$. This contradiction shows that *TRCL* does not have the bounded degree property. This proof also shows that *deterministic* transitive closure violates the BDP. (Deterministic transitive closure is defined just as transitive closure, except that one only considers paths where each node other than the last one has outdegree 1, see [29].)

What makes the BDP particularly interesting, is the following result.

Theorem 11 ([11]). *Every Gaifman-local formula has the bounded degree property.* □

From results in the previous subsection, we conclude that first-order logic and various counting logics we considered have the bounded degree property. This confirms out intuition that these logics lack mechanisms for expressing recursive (fixpoint) computation.

Corollary 6 ([32,24]). *Every Hanf-local formula has the bounded degree property. In particular,* $\mathcal{L}_{\infty\omega}(\mathbf{Q}_u)^\omega$ *formulae,* FO + COUNT *formulae (without free numerical variables), and* FO *formulae have the bounded degree property.* □

Since deterministic transitive closure does not have the BDP, we obtain the following result.

Corollary 7 ([32,24]). *Deterministic transitive closure is not definable in* $\mathcal{L}_{\infty\omega}(\mathbf{Q}_u)^\omega$ *nor in* FO + COUNT. □

This follows immediately from the BDP, and avoids all the combinatorial arguments in Section 4, and especially the ones in Section 3, that are based on game-theoretic techniques.

We now give another example that shows how the BDP can be applied to prove inexpressibility results.

Example 4. A *balanced binary tree* is a (directed) binary tree in which all paths from the root to the leaves are of the same length. Can this property be tested in FO, or perhaps in more expressive logics such as $\mathcal{L}_{\infty\omega}(\mathbf{Q}_u)^\omega$? We now use the BDP to give the negative answer.

Suppose that we have a sentence Φ that tests if a given graph is balanced binary tree. We next define a query $\varphi(x, y)$ as follows. It first defines a new graph, by interchanging the immediate successors of x, x' and x'', and the immediate successors of y, y' and y'', as shown in Figure 3 below, and then it tests is the resulting graph is a balanced binary tree. If either x or y fails to have exactly two immediate successors, then $\varphi(x, y)$ will evaluate to *false*. Assuming Φ is in the logic, so is $\varphi(x, y)$, for logics like FO and $\mathcal{L}_{\infty\omega}(\mathbf{Q}_u)^\omega$.

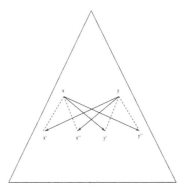

Fig. 3. Changing successors of nodes in a balanced binary tree

We now show that $\varphi(x, y)$ violates the BDP. Assume it does have the BDP, and let $N = f_\varphi(2) + 1$. Let $\mathcal{A}$ be a balanced binary tree where each path from the root to a leaf has length N. Since degrees in $\mathcal{A}$ do not exceed 2, deg_count($\varphi[\mathcal{A}]$) $< N$ by the BDP. We can see that $\mathcal{A} \models \varphi(a, b)$, for two nodes a, b, if and only if a and b are at the same level in $\mathcal{A}$. Thus, $\varphi[\mathcal{A}]$ is a disjoint union of N cliques of different sizes, and hence deg_count($\varphi[\mathcal{A}]$) $= N$. This contradiction shows that φ does not have the BDP, and hence cannot be defined in $\mathcal{L}_{\infty\omega}(\mathbf{Q}_u)^\omega$. Consequently, testing for balanced binary trees is not $\mathcal{L}_{\infty\omega}(\mathbf{Q}_u)^\omega$-definable.

One may have noticed that there is a certain asymmetry in the definition of the BDP. In the assumption, we deal with deg_set($\mathcal{A}$), but the conclusion puts a bound on deg_count($\varphi[\mathcal{A}]$). Can the definition be made symmetric? To formalize this, define the *strong bounded degree property* of $\varphi(\boldsymbol{x})$ as follows: there exists a function $f_\varphi : \mathbb{N} \to \mathbb{N}$ such that deg_count($\varphi[\mathcal{A}]$) $\leq f_\varphi$(deg_count($\mathcal{A}$)).

Proposition 2 ([11]). *There are first-order definable graph queries that violate the strong bounded degree property.* □

In fact, [11] shows that even a weaker property is violated by some first-order queries. Define the *interval bounded degree property* of a query $\varphi(\boldsymbol{x})$ as the existence of a function $f_\varphi : \mathbb{N} \to \mathbb{N}$ such that deg_count($\varphi[\mathcal{A}]$) $\leq f_\varphi(k)$ whenever deg_set($\mathcal{A}$) $\subseteq \{n, n + 1, \dots, n + k\}$ for some number n. Then there exist first-order definable queries on graphs that violate this property.

To summarize, we have seen four different locality conditions: strong Gaifman-locality, Hanf-locality, Gaifman-locality, and the bounded degree property. The relationship between them is shown in Figure 5.4. While Hanf-locality is closely tied to a game and relatively easy to show for FO and some of its extensions, Gaifman-locality and the bounded degree property are very easy to use in expressibility proofs. Fortunately, they are implied by Hanf-locality of a logic.

$$\boxed{\text{Strongly Gaifman-local} \quad \Rightarrow \quad \text{Hanf-local} \quad \Rightarrow \quad \text{Gaifman-local} \quad \Rightarrow \quad \text{BDP}}$$

Fig. 4. The relationship between the notions of locality

6   Applications in Complexity Theory

Fagin's theorem, that equates existential second-order logic and complexity class NP, started a new line of research in complexity theory. In the past 20 years, many complexity classes have been characterized in logical terms, see [13,28] for an overview. For example, polynomial time and space can be characterized by least- and partial fixpoint logics, respectively. Essential for many characterizations is the presence of a linear order on the input. The intuition behind having an order is simulating the order in which elements of the input appear on the tape of a Turing machine. While for order-invariant queries, the exact order does not affect the output, its presence is required for a logic to simulate the computation of a machine. In fact, it remains an open problem whether there is a logic for polynomial time properties of unordered graphs, for example.

In this section, we deal with a circuit complexity class TC^0. This class is defined via Boolean circuits. Consider a family of circuits $C = \{c_1, c_2, \dots, c_n, \dots\}$, where the circuit c_n has n inputs and one output. Given a Boolean string x, we say that C accepts x if the output of c_n on x is 1, whenever x is of length n.

The class AC^0 is defined as the class of languages accepted by circuits C where each gate is either an AND, or an OR, or a NOT gate, with AND and

OR gates having unbounded fan-in (no restriction on the number of inputs). The number of gates in c_n is polynomial in n, and the depth of circuits c_n is constant. (More generally, for AC^k, the depth of c_n is allowed to be $O(\log^k n)$.) The class TC^0 is defined as AC^0, except that majority gates MAJ are also allowed. Assume such a gate has k inputs. Then its single output is 1 iff at least $\lfloor k \rfloor + 1$ of its inputs are 1.

The class TC^0 is not an idle creation of complexity theory; in fact, it is of special importance in computer science. It characterizes the complexity of such important operations as integer multiplication, division, and sorting, and serves as a computational model for neural nets [44]. We refer the reader to survey [4] for additional information on circuit complexity.

Despite its importance, not much is known about the relationship between TC^0 and other complexity classes. We do know that $AC^0 \subset TC^0$, as the parity language (strings with even number of 1s) is in TC^0, but not in AC^0 [8]. We also know (see [4], for example) that

$$TC^0 \subseteq NC^1 \subseteq DLOG \subseteq NLOG \subseteq NC \subseteq PTIME \subseteq NP,$$

but we do not know if *any* of the inclusions is proper! In fact, [45] showed that there is inherent difficulty in separating TC^0 from NP, at least using conventional techniques of circuit complexity. A general notion of *natural proof* was formulated in [45]; this notion subsumes most of the existing lower bound proofs. Then [45] showed that the existence of a natural proof separating these two classes would imply that no good pseudo-random number generators are computable in TC^0. Putting it in the language of [4], it would imply that no cryptographically-secure functions can be computed in TC^0, even though cryptographers believe that such functions do exist. As the notion of natural proof is quite different from results on logical expressibility in finite model theory, one might attempt to avoid obstacles of [45] by using a logical characterization of TC^0. Below, we survey results in this direction.

Notice that in the definition of classes AC^0 and TC^0, we did not say anything about the relationship between circuits $c_n \in C$ when n varies; in fact, they can compute completely "different" things for different n. However, in most applications, those circuits compute the same property, like parity. Capturing this intuition leads to the notion of *uniformity*. The weakest notion of uniformity is PTIME-uniformity, meaning that the mapping $n \mapsto c_n$ is computable in polynomial time. Similarly, one can define logspace-uniformity (see [3] for using these notions with the class TC^0). However, the most widely used notion of uniformity is DLOGTIME-uniformity. We spare the reader the more technical definition, that can be found in [5], and instead give the characterization theorem.

Theorem 12 ([5]). *DLOGTIME-uniform* TC^0 $=$ FO + COUNT+ $<$.   □

From now on, when we speak of TC^0, we mean its DLOGTIME-uniform version; that is, FO + COUNT+ $<$. The latter is the class of problems definable by FO+COUNT formulae in the presence of an order relation $<$. We will mostly deal with order-independent properties. The notion of order-independence is defined as follows. Suppose we have a FO + COUNT formula φ in the language of σ and $<$, and suppose $\mathcal{A}$ is a σ-structure. Then, for any two orderings $<_1$ and $<_2$ on $\mathcal{A}$, and for any a, $\mathcal{A}_{<_1} \models \varphi(a)$ iff $\mathcal{A}_{<_2} \models \varphi(a)$, where $\mathcal{A}_<$ denotes the extension of $\mathcal{A}$ with the order $<$.

Even though we restrict our attention to order-independent properties, the mere presence of an order relation does increase the expressive power:

Proposition 3 ([6]). *There are order-independent properties definable in* FO + COUNT+ $<$ *but not in* FO + COUNT. $\qquad \square$

The example of a separating query (not the proof!) is quite simple. Let $\sigma = (E, U)$, where E is binary and U is unary. Consider the following property: If the interpretation of E is an equivalence relation, then the number of distinct sizes of equivalence classes of E equals the cardinality of U. This query is not definable in FO + COUNT [6], but, as shown in [38], can easily be defined with order, since all elements whose equivalence classes have the same size, can be canonically represented by the $<$-minimal such element. Then one just checks if the number of those elements equals the cardinality of U.

Theorem 12 reduces the problem of separating TC^0 from classes above it to the problem of logical expressibility; for example, to show $\mathrm{TC}^0 \neq \mathrm{NLOG}$, it would suffice to show that transitive closure, an NLOG-complete problem, is not definable in FO + COUNT+ $<$. Since locality gives us an easy proof that transitive closure is not in FO + COUNT, one might try to push the ideas of locality into the ordered setting.

We do not know whether the above expressivity bound on FO + COUNT+ $<$ is true, although we conjecture that it is. Below, we survey some of the partial results confirming the intuition. We state the results for the NLOG-complete problem of computing the transitive closure, but they also hold for *deterministic* transitive closure, which is complete for DLOG.

Assume that instead of an order relation, we have a *successor* relation SUCC. Since it realizes only degrees 0 and 1, as an immediate consequence of the bounded degree property of FO + COUNT, we obtain

Corollary 8 ([14]). *Transitive closure is not definable in* FO + COUNT + SUCC. $\qquad \square$

Note that FO plus transitive closure *TRCL* plus successor relation capture NLOG (cf. [13,28]); hence, FO + COUNT + SUCC $\subset$ NLOG = FO + *TRCL* + SUCC. This result was first shown in [14], via a rather complex argument based on games of [30]. Later, using the results of [40], the journal version of [14] (see [15]) gave a much simpler proof based on Hanf's condition. Finally, using the bounded degree property, we gave a completely elementary proof.

The use of bounded degree property allows us to substitute any auxiliary relation for SUCC, as long as its degrees are bounded by a constant. For example, we could use balanced binary trees (note that using such a structure would most certainly make a game-based proof unmanageable). The next question is: How can we lift the results for FO + COUNT from the constant world to that where degrees are allowed to depend on the size of a structure?

First such result was given in [32], and it used the notion of *moderate* degree of [17]. Let $\mathcal{C}$ be a class of structures. Let $maxdeg_{\mathcal{C}}(n)$ denote the maximal degree of a structure in $\mathcal{C}$, whose cardinality is n. Then we say that $\mathcal{C}$ is a class of relations of moderate degree if $maxdeg_{\mathcal{C}}(n) \leq \log^{o(1)} n$. That is, for some function $\delta : \mathbb{N} \to \mathbb{N}$ with $\lim_{n\to\infty} \delta(n) = 0$, we have $maxdeg_{\mathcal{C}}(n) \leq \log^{\delta(n)} n$. Combination of results from [11] and [32] led to the following.

Proposition 4 ([32]). *Transitive closure is not definable in* FO + COUNT *in the presence of relations of moderate degree.*   □

A linear order on an n-element set realizes n different degrees, from 0 to $n-1$. Thus, we need to lift the results from relations of small (constant or moderate) degree to relations of large (comparable with the size of the input) degree. The concept of moderate degree was introduced in [17] to show that connectivity is not definable in monadic Σ_1^1 in the presence of those relations. Later, [46] extended this to linear orders. Thus, one may ask if a similar avenue of attack on the separation problem can be pursued in the case of FO + COUNT.

A partial result in this direction exists. Let $\mathcal{O}_k$ stand for the class of relations which are pre-orders $\langle A, \prec \rangle$ (i.e., $\prec$ is reflexive and transitive), and each equivalence class of the relation $x \sim y \equiv (x \prec y) \wedge (y \prec x)$ has at most k elements. In particular, $\mathcal{O}_1$ is the class of linear orders.

Theorem 13 ([38]). *Transitive closure is not definable in* FO + COUNT + $\mathcal{O}_k$, *for any* $k > 1$.   □

In fact, the result of [38] is even stronger as it allows relations from $\mathcal{O}_2$ to be of special form: those that have at most $g(n)$ elements in equivalence classes of size 2, and the rest, $n - g(n)$ elements, form a linear order. The function $g(n)$ can be chosen to be arbitrarily small, e.g., $\log \log \ldots \log n$, but it cannot be bounded by a constant. The version of Theorem 13, proved in [38] for deterministic transitive closure, implies DLOG $\not\subseteq$ FO + COUNT + $\mathcal{O}_k$ for any $k > 1$. But this still falls short of resolving the most important case of $k = 1$.

One of the main goals of pursuing this line of research was to avoid the obstacles on the path towards separating TC^0 from other classes posed by the natural proofs of [45]. At the first glance, our expressivity bounds look nothing like natural proofs, although we must admit that there was no systematic study conducted on the relationship between classical lower bounds proofs in circuit complexity and logical expressivity bounds for FO + COUNT and the likes.

However, while avoiding one kind of problems, we encountered (perhaps even created) different ones. For example, the proof of Theorem 13 is based on the following technique. One first shows that if for a class $\mathcal{C}$ of auxiliary relations, there exists a class of graphs $\mathcal{G}$ such that the pair $(\mathcal{C}, \mathcal{G})$ satisfies two properties, P1 and P2, then transitive closure of graphs in $\mathcal{G}$ is not definable in FO + COUNT + $\mathcal{C}$. Then, [38] showed how to construct $\mathcal{G}$ for $\mathcal{C}$ being $\mathcal{O}_k$, with $k > 1$, thus proving inexpressibility of transitive closure in FO + COUNT + $\mathcal{O}_k$.

Unfortunately, [38] also showed that if $C = \mathcal{O}_1$, the class of linear orders, then there is *no* class of graphs $\mathcal{G}$ such that $(C, \mathcal{G})$ satisfies P1 and P2. That is, the approach is inherently limited for showing results in the ordered setting.

Another large obstacle to using the ideals of locality for proving the separation is a recent result:

Theorem 14 ([24]). *There exist order-invariant formulae in* FO+COUNT+ $<$ *that do not have the bounded degree property. Consequently, order-invariant* FO+ COUNT+ $<$ *is neither Gaifman-local nor Hanf-local.*   □

Nevertheless, we believe it was a useful exercise to study the expressivity of FO + COUNT with relations of large degree, viewing it as a reasonable approximation of uniform TC^0. The bounds of [14], obtained less than 4 years ago with a great deal of effort, were shown here by a simple application of the bounded degree property. The best lower bound given by locality arguments (Theorem 13) shows that one can get very close to a linear order, and thus very close to uniform TC^0, without achieving the power of logspace computation. In a way, results of this section can be viewed as separating "very uniform" versions of TC^0 from DLOG and classes above it.

7 Applications in Database Theory

The theory of database query languages is firmly grounded in finite-model theory, and also provides a major motivation for finite-model theory research. Traditional databases query languages, such as relational algebra and calculus, have precisely the power of first-order logic. An important subclass, called conjunctive queries, is simply the $\{\exists, \wedge\}$-fragment of first-order logic. Various extensions of relational calculus, such as Datalog, Datalog with negation, and the language of *while* loops, correspond to various fixpoint extensions of first-order logic, see [1].

In the relational model, data is stored in relations. For example, a database may have two relations, for storing information about employees (called Emp) and departments (called Dept). Assume that Emp stores triples containing employee name, department, and salary, and Dept stores triples containing department name, manager, and manager's salary. Below is an example of a database:

EName	EDept	ESalary
John	A1	50
Ann	A1	60
Jim	B2	75

DName	Manager	MSalary
A1	Bob	80
B2	Steve	85
C3	Mary	80

An example of a query is *"For each employee, find his or her manager."* This can be written in first-order logic as:

$$q(emp, manager) \equiv$$
$$\exists dept, esal, msal \; \mathtt{Emp}(emp, dept, esal) \wedge \mathtt{Dept}(dept, manager, msal).$$

In real life, of course, programmers do not write first-order formulae; instead they write queries in the language called SQL, which is the *lingua franca* of the commercial database world. The above query, in SQL, will look like:

```
SELECT E.EName, D.Manager
FROM   Emp E, Dept D
WHERE  E.Edept = D.Dname
```

The best way to read such statements is as set-theoretic comprehensions: the above becomes

$$
\left\{ (EName,\ D.Manager) \ \middle|\ \begin{array}{l} (EName, EDept, ESalary) \in \texttt{Emp}, \\ (DName, Manager, MSalary) \in \texttt{Dept}, \\ EDept = DName \end{array} \right\}
$$

The basis of SQL is the select-from-where statement, with the addition union and difference, and features such as view creation, which make the language compositional. This basis has precisely the power of the first-order logic. However, all practical implementations of SQL come equipped with two additional features: *arithmetic operations*, and *aggregate functions*. For example, consider the following query: *"Find all departments that have more than 5 employees, together with the name of the manager and the average salary of the employees."*

In SQL, this will be written as

```
SELECT D.DName, D.Manager, AVG(E.ESalary)
FROM   Emp E, Dept D
WHERE  E.Edept = D.Dname
GROUPBY D.DName
HAVING COUNT(E.EName) > 5
```

There are two key new features in this query. *Grouping* is given by the clauses GROUPBY and HAVING: for each department manager, we group together all the employees in his/her department, provided there are more than five of them. The other feature is *aggregate function*: these are AVG, for computing the average salary, and COUNT, for counting the number of employees.

Let us now see why adding these features is indeed an extension of first-order logic. We consider a query that we know is inexpressible in first-order logic alone: given a graph G, find the set of nodes x with in-deg(x) = out-deg(x). We let graphs be represented as a binary relation Edges with two attributes, From and To. As the first step, we create two new relations, one storing nodes together with their in-degrees, and the other storing nodes with their out-degrees. Such intermediate relations are called *views* in SQL, and are created as

```
CREATE VIEW INDEGV(Node, Indeg) AS
  SELECT  E.To, COUNT(E.From)
  FROM    Edges E
  GROUPBY E.To
```

```
CREATE VIEW OUTDEGV(Node, Outdeg) AS
  SELECT  E.From, COUNT(E.To)
  FROM    Edges E
  GROUPBY E.From
```

Using these views, one computes the answer to the query as

```
SELECT INDEGV.Node
FROM   INDEGV, OUTDEGV
WHERE  (INDEGV.Node = OUTDEGV.Node)
       AND (INDEGV.Indeg = OUTDEGV.Outdeg)
```

By now, the reader must be convince that Härtig and Rescher quantifiers can be expressed in SQL. Thus, it is more powerful than first-order logic. The question now is:

How expressive is SQL?

More precisely, there seems to be a "folk result" saying that SQL cannot compute recursive queries, such as transitive closure. That is, it lacks a mechanism for recursive computation. The question now is: What kind of formal statement can one prove to confirm this intuition?

The approach of some textbooks is to restrict SQL to its subset which is essentially first-order logic, and use expressivity bounds for the latter. However, as we have just seen, this is not satisfactory. The difficulty with answering the question above is that there are dozens of different version of SQL (see [43] for an overview of standards and dialects), and they often support different sets of operators. For example, some versions even add the transitive closure operator. Thus, we restrict our attention to features that are common to *all* versions of SQL; that is, grouping and aggregation (cf. [1,43]). The first result on the expressive power of such a language was based on yet unproven assumption from complexity theory, and two observations: SQL queries can be evaluated in deterministic logspace, and transitive closure is complete for nondeterministic logspace. Thus,

Proposition 5 ([10]). *Assume that $DLOGSPACE \neq NLOGSPACE$. Then transitive closure cannot be expressed in SQL.*   □

Can we get rid of the unproven assumption? The problem we face is that SQL per se is quite inconvenient to work with – its syntax is quite awkward, and in fact it has been an object of persistent criticism. SQL combines sets and multisets in order to evaluate aggregates: for example, computing the average salary, one *cannot* first project out the salary attribute and then compute its average, as the elimination of duplicates will produce an incorrect result. Following [20,35], a language that deals correctly with multiset and set semantics was proposed in [36]. As this language can model the main features of SQL, and extends first-order logic, it was suggested to use it as a rational reconstruction of SQL.

Let us now give an informal introduction into this language, which we call here AGGR$_\mathbb{Q}$. It deals with objects which can freely combine rational numbers, elements of the domain of atomic values, D, tuples, and sets. In particular, it permits sets of sets. The language is *statically typed* (cf. [21,36]), but we shall not go into detail here. The full description can be found in [21,36]; here we highlight the salient features. The language allows one to apply a function to each element of a set, that is, obtain a set $\{f(x) \mid x \in X\}$ from X. If all elements of a set $\mathcal{X}$ are sets themselves, their union can be taken: $\bigcup_{X \in \mathcal{X}} X$. Basic arithmetic operations $(+, -, *, \div, <)$ are available on $\mathbb{Q}$. On the domain of basic values D, only equality test is available (notice the absence of order). Finally, if f is a function into $\mathbb{Q}$, its values on a set X can be added up; that is, one can compute $\sum_{x \in X} f(x)$. Note that if f is identically 1, then the above is the cardinality of X.

It was proved in [36] that these features model the main features of SQL. Furthermore, extending a result from [49], a conservativity property was shown in [36]. It says that nesting of sets is in a sense superfluous: every query from relational databases to relational databases in AGGR$_\mathbb{Q}$ can be written in a way that does not use sets of sets. Nesting is essential for modeling grouping, and thus the (nontrivial) conservativity result gives us a rather pleasant language (without higher-order features) to model *all* the features of SQL. In [36], the following was proved:

Proposition 6 ([36]). *Transitive closure cannot be expressed in* AGGR$_\mathbb{Q}$.   □

While this does provide useful bounds, there are two problems with the proof in [36]. First, it is very syntactic. It proceeds by establishing a normal form result for queries on a special class of inputs; the property depends both on the class of inputs, and the properties of the chosen syntax of the language. Thus, making a minor change in the syntax that does not affect expressiveness would mean that the proof must be redone from scratch. Even more unpleasantly, the proof in [36] establishes inexpressibility of transitive closure, but fails to establish a general property that will give us expressivity bounds.

In [21,36] it was conjectured that *relational queries* in AGGR$_\mathbb{Q}$ have the BDP. By relational queries we mean those whose inputs and outputs only contain elements of D, but no numbers, although rational numbers can be used in the process of evaluating a query. An SQL example of finding nodes with equal in- and out-degrees shows that there are relational queries definable in SQL but not in first-order. Clearly, proving the above conjecture would resolve the problem for queries such as transitive closure or deterministic transitive closure. By extending the normal form result AGGR$_\mathbb{Q}$, the following was proved.

Proposition 7 ([11]). *Every relational query in* AGGR$_\mathbb{Q}$ *has the bounded degree property.*   □

It is still unpleasant that the proof depends on a particular syntax for the language. Also, it would be interesting to know if relational queries are Gaifman-local. These two questions were considered in [32]. The approach taken by [32] was the following. Restrict the language to *natural numbers* only. That is, one can compute COUNT but not AVG. Let us call this language AGGR$_\mathbb{N}$. With rationals out of the way, try to embed it into FO + COUNT to prove locality.

Embedding into FO + COUNT turns out to be problematic, as in FO + COUNT the sizes of the first-sort and second-sort universes are the same. In contrast, in SQL, one can create numbers much bigger than the size of the database. A simple example is this: consider an n-element set X, and compute $\sum_{x \in X} f(x)$ where $f(x) = 2$ for every x. The result is $2n$. The solution to the problem is to modify a query, essentially by putting a huge linear order "on the side" and having all arithmetic done on that linear order. This technique led to the following result.

Proposition 8 ([32]). *For any relational query Q in AGGR$_\mathbb{N}$, another query Q' can be found such that (a) Q is Gaifman-local iff Q' is, and (b) Q' can be defined in FO + COUNT. Consequently, every relational query in AGGR$_\mathbb{N}$ is Gaifman-local.* □

Using this technique, [37] returned to the main question: what is the expressive power of SQL? It defined a new language, called AGGR$_\mathbb{N}^{\text{flat}}$, which first restricts AGGR$_\mathbb{N}$ to objects that do not contain sets of sets, and then adds new arithmetic operation, and *product* over a set. That is, given a set X and a function f into $\mathbb{N}$ definable in AGGR$_\mathbb{N}^{\text{flat}}$, one can compute $\prod_{x \in X} f(x)$ in AGGR$_\mathbb{N}^{\text{flat}}$. Then the following sequence of results was proved.

Theorem 15 ([37]). *1) Every relational AGGR$_\mathbb{Q}$ query can be expressed in the language AGGR$_\mathbb{N}^{\text{flat}}$.*
2) For any relational query Q in AGGR$_\mathbb{N}^{\text{flat}}$, another query Q' in AGGR$_\mathbb{N}^{\text{flat}}$ can be found such that (a) Q is Gaifman-local iff Q' is, and (b) Q' can be defined in FO + COUNT. □

The bulk of the proof of this theorem is in showing (b) of part 2). Combining these results with locality of FO + COUNT, we obtain:

Corollary 9. *Every relational query in AGGR$_\mathbb{Q}$ is Gaifman-local.* □

Consequently, plain SQL queries are Gaifman-local and have the bounded degree property. Therefore, queries that need a recursion mechanism (transitive closure, deterministic transitive closure, connectivity test, etc.) *cannot* be computed in SQL.

Thus, locality helped us answer important questions about expressive power of real world database query languages.

We conclude this section by a remark about the set of basic operations on the domain of atomic values, D. In all the above results, we assume that it is only possible to test if two elements of D are equal. In many applications, there is a meaningful linear order on such domains (for examples, salaries can be compared). What happens if D is ordered? As it often happens, things become a lot more complicated in the ordered setting. In a way, we saw it in the last section. For SQL, one can show that it is possible to express any query from

FO + COUNT+ $<$ if D is ordered. Thus, every uniform TC^0 property becomes definable in SQL, and hence the problems of expressivity in of recursive queries (such as transitive closure) hinges on the separation of complexity classes, thus confirming the original intuition of [10]. Note that there are other well-known examples of close connection between separation of complexity classes and expressivity bounds of query languages, see [2].

One can go further and add more operations to D. For example, one can assume that D is the field of real numbers with the usual operations $+, *, -$, or perhaps more complex such as e^x. The results of [7] show that for the class of *generic* [1] queries (those that commute with permutations of D), these extra operations do not add expressive power, beyond a linear order. Most examples of queries we consider – transitive closure, connectivity test, etc. – are generic. Thus, adding extra operations beyond $<$ does not lead to an increase in power of languages such as first-order logic or SQL. This assertion, in the case of an interpreted domain D, depends on what one means by quantification $\exists x$. In most application, quantification in queries assumes the finite database; however, it is conceivable (and in some applications, important) to quantify over D. We refer the reader to [7] for the discussion on this topic.

8   Conclusion

In this paper, we reviewed some results that were developed for proving lower bounds for logical expressibility. We considered first-order logic and its extensions with several kinds of counting mechanisms. We presented the usual game-theoretic characterizations of those logics, as well as Gaifman's and Hanf's theorems, and general notions behind these results. We also studied the relationship between these notions and the bounded degree property. We reviewed applications of these notions in descriptive complexity theory and database theory.

Note. Several results on locality of logics appeared recently. As it is too late to include their detailed descriptions in this survey, we just give pointers to relevant papers. Most importantly, [19] partly extended locality techniques to the ordered setting. Namely, [19] proves that every order-invariant query definable in FO+ $<$ is Gaifman-local, where the concept of neighborhood does not take into account the order relation. An elementary proof of Gaifman-locality of a very powerful counting logic was given in [33]; that paper also used a refined argument for proving locality to find tight bounds on the radii of neighborhoods required in locality theorems. Here, we typically used the radii of the order $O(3^{qr(\varphi)})$; [33] improves this to $O(2^{qr(\varphi)})$. Hanf-locality of the counting logic from [33] was proved in [25]. It was also shown in [25] that Hanf-locality remains valid under the addition of aggregate functions, which provided a transparent proof of Corollary 9. Finally, [34] improved Theorem 13 by showing that invariant queries in FO + COUNT + $\mathcal{O}_k$, $k > 1$, have the bounded degree property. This remains true for more powerful logics, and a more general class of auxiliary relations.

Acknowledgement

We thank Jouko Väänänen for the invitation to write this survey. We are grateful to Lauri Hella and Limsoon Wong for their comments.

References

1. S. Abiteboul, R. Hull and V. Vianu. *Foundations of Databases*, Addison Wesley, 1995.
2. S. Abiteboul and V. Vianu. Computing with first-order logic. *Journal of Computer and System Sciences* 50 (1995), 309–335.
3. M. Agrawal, E. Allender and S. Datta. On TC^0, AC^0, and arithmetic circuits. In *Proc. 12th IEEE Conf. on Computational Complexity*, 1997.
4. E. Allender. Circuit complexity before the dawn of the new millennium. In *Proc. 16th Conf. on Foundations of Software Technology and Theoretical Computer Science (FST&TCS'96)*, Springer LNCS vol. 1180, 1996, 1–18.
5. D.A. Barrington, N. Immerman and H. Straubing. On uniformity within NC^1. *J. Comput. and Syst. Sci.*, 41:274–306,1990.
6. M. Benedikt, H.J. Keisler. On expressive power of unary counters. *Proc. Intl. Conf. on Database Theory (ICDT'97)*, Springer LNCS 1186, January 1997.
7. M. Benedikt and L. Libkin. Relational queries over interpreted structures. *J. ACM*, to appear. Extended abstract in *PODS'97*, pages 87–98.
8. R.B. Boppana and M. Sipser. The Complexity of Finite Functions. In *"Handbook of Theoretical Computer Science,"* Volume A, Chapter 14, pages 759–804, North Holland, 1990.
9. J. Cai, M. Fürer and N. Immerman. On optimal lower bound on the number of variables for graph identification. *Combinatorica*, 12 (1992), 389–410.
10. M. Consens and A. Mendelzon. Low complexity aggregation in GraphLog and Datalog, *Theoretical Computer Science* 116 (1993), 95–116. Extended abstract in *ICDT'90*.
11. G. Dong, L. Libkin and L. Wong. Local properties of query languages. *Theoretical Computer Science*, to appear. Extended abstract in *ICDT'97*, LNCS vol. 1186, pages 140–154.
12. H.-D. Ebbinghaus. Extended logics: the general framework. In J. Barwise and S. Feferman, editors, *Model-Theoretic Logics*, Springer-Verlag, 1985, pages 25–76.
13. H.-D. Ebbinghaus and J. Flum. *Finite Model Theory*. Springer Verlag, 1995.
14. K. Etessami. Counting quantifiers, successor relations, and logarithmic space, In *Proc. Structure in Complexity Theory*, 1995.
15. K. Etessami. Counting quantifiers, successor relations, and logarithmic space, *Journal of Computer and System Sciences*, 54 (1997), 400–411.
16. R. Fagin. Easier ways to win logical games. In *Proc. DIMACS Workshop on Descriptive Complexity and Finite Models*, AMS 1997.
17. R. Fagin, L. Stockmeyer and M. Vardi, On monadic NP vs monadic co-NP, *Information and Computation*, 120 (1994), 78–92.
18. H. Gaifman. On local and non-local properties, *Proceedings of the Herbrand Symposium, Logic Colloquium '81*, North Holland, 1982.
19. M. Grohe and T. Schwentick. Locality of order-invariant first-order formulas. In *MFCS'98*, pages 437–445.
20. S Grumbach and T. Milo. Towards tractable algebras for bags. *Journal of Computer and System Sciences*, 52 (1996), 570–588.

21. S. Grumbach, L. Libkin, T. Milo and L. Wong. Query languages for bags: expressive power and complexity. *SIGACT News*, 27 (1996), 30–37.
22. W. Hanf. Model-theoretic methods in the study of elementary logic. In J.W. Addison et al, eds, *The Theory of Models*, North Holland, 1965, pages 132–145.
23. L. Hella. Logical hierarchies in PTIME. *Information and Computation*, 129 (1996), 1–19.
24. L. Hella, L. Libkin and J. Nurmonen. Notions of locality and their logical characterizations over finite models. *Journal of Symbolic Logic*, to appear.
25. L. Hella, L. Libkin, J. Nurmonen and L. Wong. Logics with aggregate operators. In *Proc. 14th IEEE Symp. on Logic in Computer Science (LICS'99)*, Trento, Italy, July 1999, pages 35–44.
26. L. Hella and J. Nurmonen. Vectorization hierarchies of some graph quantifiers. *Archive for Mathematical Logic*, to appear.
27. L. Hella and G. Sandu. Partially ordered connectives and finite graphs. In M. Krynicki, M. Mostowski and L. Szczerba, eds, *Quantifiers: Logics, Models and Computation II*, Kluwer Academic Publishers, 1995, pages 79–88.
28. N. Immerman. *Descriptive Complexity*. Springer-Varlag, 1999.
29. N. Immerman. Languages that capture complexity classes. *SIAM J. Comput.* 16 (1987), 760–778.
30. N. Immerman and E. Lander. Describing graphs: A first order approach to graph canonization. In *"Complexity Theory Retrospective"*, Springer Verlag, Berlin, 1990.
31. Ph. Kolaitis and J. Väänänen. Generalized quantifiers and pebble games on finite structures. *Annals of Pure and Applied Logic*, 74 (1995), 23–75.
32. L. Libkin. On the forms of locality over finite models. In *Proc. 12th IEEE Symp. on Logic in Computer Science (LICS'97)*, Warsaw, Poland, June–July 1997, pages 204–215.
33. L. Libkin. On counting logics and local properties. In *Proc. 13th IEEE Symp. on Logic in Computer Science (LICS'98)*, Indianapolis, June 1998, pages 501–512.
34. L. Libkin. Logics with counting, auxiliary relations, and lower bounds for invariant queries. In *Proc. 14th IEEE Symp. on Logic in Computer Science (LICS'99)*, Trento, Italy, July 1999, pages 316–325.
35. L. Libkin and L. Wong. Some properties of query languages for bags. In *Proc. Database Programming Languages 1993*, Springer Verlag, 1994, pages 97–114.
36. L. Libkin and L. Wong. Query languages for bags and aggregate functions. *Journal of Computer and System Sciences* 55 (1997), 241–272. Extended abstract in *PODS'94*, pages 155–166.
37. L. Libkin and L. Wong. On the power of aggregation in relational query languages. In *Proc. Database Programming Languages 1997*, Springer LNCS 1369, pages 260–280.
38. L. Libkin and L. Wong. Unary quantifiers, transitive closure, and relations of large degree. In *Proc. 15th Symp. on Theoretical Aspects of Computer Science (STACS'98)*, Springer LNCS 1373, pages 183–193.
39. K. Luosto. Hierarchies of monadic generalized quantifiers. *Journal of Symbolic Logic*, to appear.
40. J. Nurmonen. On winning strategies with unary quantifiers. *J. Logic and Computation*, 6 (1996), 779–798.
41. J. Nurmonen. Counting modulo quantifiers on finite structures. *Information and Computation*, to appear. Extended abstract in *Proc. 11th IEEE Symp. on Logic in Computer Science (LICS'96)*, New Brunswick, NJ, July 1996, pages 484–493.
42. J. Nurmonen. Unary quantifiers and finite structures. PhD Thesis, University of Helsinki, 1996.

43. P. O'Neil. *Database*. Morgan Kaufmann, 1994.
44. I. Parberry and G. Schnitger. Parallel computation and threshold functions. *Journal of Computer and System Sciences* 36 (1988), 278–302.
45. A. Razborov and S. Rudich. Natural proofs. *Journal of Computer and System Sciences* 55 (1997), 24–35.
46. T. Schwentick. On winning Ehrenfeucht games and monadic NP. *Annals of Pure and Applied Logic*, 79 (1996), 61–92.
47. T. Schwentick and K. Barthelmann. Local normal forms for first-order logic with applications to games and automata. In *Proc. 15th Symp. on Theoretical Aspects of Computer Science (STACS'98)*, Springer LNCS 1373, pages 444-454.
48. J. Väänänen. Generalized quantifiers, an Introduction. This volume.
49. L. Wong. Normal forms and conservative properties for query languages over collection types. *Journal of Computer and System Sciences* 52 (1996), 495–505. Extended abstract in *PODS 93*.

A Perspective on
Lindström Quantifiers and Oracles

Iain A. Stewart*

Department of Mathematics and Computer Science, Leicester University
Leicester LE1 7RH, U.K.
i.a.stewart@mcs.le.ac.uk
http://www.mcs.le.ac.uk/ĩstewart

Abstract. This paper presents a perspective on the relationship between Lindström quantifiers in model theory and oracle computations in complexity theory. We do not study this relationship here in full generality (indeed, there is much more work to do in order to obtain a full appreciation), but instead we examine what amounts to a thread of research in this topic running from the motivating results, concerning logical characterizations of nondeterministic polynomial-time, to the consideration of Lindström quantifiers as oracles, and through to the study of some naturally arising questions (and subsequent answers). Our presentation follows the chronological progress of the thread and highlights some important techniques and results at the interface between finite model theory and computational complexity theory.

1 Introduction

Prior to about 1974, the model theory of finite structures had been deemed to be rather uninteresting, essentially because many of the fundamental results of (infinite) model theory, such as the Completeness and Compactness Theorems, fail when only finite structures are allowed (see [17]). This is not to say that there are no interesting results in the finite case: Trakhtenbrot's Theorem [38] and the 0-1 law for first-order logic [12] are two. But such results appeared sporadically and there was no real concerted research effort in this regard. However, all this changed thanks to Fagin's Theorem [10] which tied together finite model theory and computational complexity theory. Since then, finite model theory has witnessed an explosive growth.

Notwithstanding the beauty of Fagin's Theorem, finite model theory still remained relatively dormant until the late seventies and early eighties. Instrumental in those early days was the work of Immerman which provided further links between finite model theory and complexity theory; and, in particular, yielded logical characterizations of numerous complexity classes by extensions of first-order logic using "operators corresponding to problems" (other operators of a very different syntactic nature, such as fixed-point operators, were

---

* Supported by EPSRC Grant GR/K 96564.

J. Väänänen (Ed.): ESSLLI'97, LNCS 1754, pp. 51–71, 1999.
© Springer-Verlag Berlin Heidelberg 1999

also used to extend first-order logic [20,39]). These operators are essentially infinite uniform sequences of Lindström quantifiers: Lindström [24,25] introduced these quantifiers in model theory so as to characterize minimal logics expressing certain properties. Immerman's logical characterizations of complexity classes yielded other things, such as new notions of (complexity-theoretic) reductions and strong (complexity-theoretic) completeness results, and his general approach enabled him to solve a long-standing open question; namely, he showed that the complexity class **NL** is closed under complementation [22] (this problem was solved independently at about the same time by Szelepcsényi [37]).

Whilst Immerman logically characterized complexity classes such as **L**, **NL** and **P** using operators of the form alluded to above, **NP** remained to be so characterized. Such a characterization was explicitly established by Stewart [31]; although unbeknownst to him, it had (essentially) already been obtained by Dahlhaus [5]. It is this logical characterization of **NP** that forms the starting point for the research travels in this paper. In more detail, we intend to show how this logical characterization of **NP** leads to the consideration of Lindström quantifiers as oracles[1] (even though there are very definite differences between the two concepts), and to a logical characterization of the complexity class $\mathbf{L}^{\mathbf{NP}}$. We then examine this characterization in more general settings in two ways: first, we consider analogous logical characterizations of complexity classes of the form $\mathbf{L}^{\mathbf{CC}}$ where **CC** is not necessarily **NP** (that is, we explore the extent to which the methods and techniques involved in the logical characterization of $\mathbf{L}^{\mathbf{NP}}$ are in some sense generic); and, second, we consider whether our logical characterization of $\mathbf{L}^{\mathbf{NP}}$, which happens to be on the class of ordered structures, holds on the class of all structures.

It is not our intention here to provide a systematic and detailed framework within which different oracle access mechanisms for oracle machines of varying resource bounds can be related to the use of Lindström quantifiers in different logics: one such framework has been laid out in [26,27]. What we aim to do is to show how logical characterizations of **NP** give rise to the study of the relationship between resource-bounded oracle computations and logics involving Lindström quantifiers. Note the wording of the previous sentence: the motivating spark for the research highlighted within this paper is essentially of the form: "*We have a logical characterization of* **NP** *as a fragment of a particular logic. So which natural complexity class, if any, is captured by the full logic?*". That is, the research arose because we wanted to find out which complexity class was captured by a certain logic, and not because we decided to model resource-bounded oracle computations in a logical fashion.

The survey presented herein highlights results in the papers [7,15,32,33]. Whilst we acknowledge that survey papers generally include explicit definitions of all concepts and notions occurring, we occasionally fail to be as detailed as we might be both for reasons of space and because it is really on the results in

---

[1] We do not mean to imply that this logical characterization of **NP** inspired the first consideration of Lindström quantifiers as oracles: such a consideration was made by Grädel [13].

the aforementioned papers that we wish to focus. There are already a number of surveys in which any such material missing from this paper might be found (notably [28]), as well as the standard finite model theory reference [9]. We assume that the reader is acquainted with the basic notions of complexity theory (as can be found in, for example, [11]). We include sketches of important proofs and we hope that we give a flavour of how current research at the interface between finite model theory and computational complexity theory is progressing. We also pose a number of problems which we believe to be worthy of investigation.

We begin, in Section 2, by giving some basic definitions relevant to the logical characterization of complexity classes before saying more about Lindström quantifiers and how they compare with oracles in Section 3. We exhibit a logical characterization of the complexity class $\mathbf{L}^{\mathbf{NP}}$, on ordered structures, in Section 4, and in Section 5, we examine the logical characterization of other log-space oracle complexity classes. In Section 6, we return to our logical characterization of $\mathbf{L}^{\mathbf{NP}}$ and consider it on the class of all structures.

2   Some Basic Definitions

We begin by giving some very basic definitions relating to how to consider a complexity class as a class of problems, where by "problem" we mean a set of finite structures (see below). The reader is encouraged to consult [9] and [11] for further details regarding finite model theory and complexity theory, respectively.

In general, a *signature* $\sigma = \langle \underline{R}_1, \underline{R}_2, \dots, \underline{R}_r, \underline{C}_1, \underline{C}_2, \dots, \underline{C}_c \rangle$ is a tuple of *relation symbols* $\{\underline{R}_i : i = 1, 2, \dots, r\}$, with $\underline{R}_i$ of arity a_i, and *constant symbols* $\{\underline{C}_i : i = 1, 2, \dots, c\}$. A *(finite) structure* of *size* n over σ is a tuple $\mathcal{A} = \langle \{0, 1, \dots, n-1\}, R_1, R_2, \dots, R_r, C_1, C_2, \dots, C_c \rangle$ consisting of a *universe* $|\mathcal{A}| = \{0, 1, \dots, n-1\}$, *relations* $R_1, R_2, \dots, R_r$ on the universe $|\mathcal{A}|$ of arities $a_1, a_2, \dots, a_r$, respectively, and *constants* $C_1, C_2, \dots, C_c$ from the universe $|\mathcal{A}|$. The size of $\mathcal{A}$ is also denoted by $|\mathcal{A}|$. We denote the set of all (finite) structures over σ by STRUCT(σ) (henceforth, we do not distinguish between relations (resp. constants) and relation (resp. constant) symbols, and we assume that all structures are finite and of size at least 2). A *problem* over σ is a subset of STRUCT(σ) which is closed under isomorphism. If Ω is some problem then the signature of Ω is denoted $\sigma(\Omega)$.

We tie together classes of problems and complexity classes as follows. Complexity classes are generally considered to be classes of languages over the alphabet $\{0, 1\}$, recognized by various resource-bounded computing devices (for example, by log-space deterministic Turing machines), whereas problems are (isomorphism closed) sets of finite structures. What we do is to encode, via some (reasonable) *encoding scheme* (see below), our structures so that they might be input to some computational device. For example, suppose that we have some structure $\mathcal{A}$ of size n in which there is a relation R of arity a. We could encode this relation as a string of n^a bits

$$R(0, 0, \dots, 0, 0), R(0, 0, \dots, 0, 1), \dots, R(0, 0, \dots, 0, n-1),$$
$$R(0, 0, \dots, 1, 0), \dots, R(n-1, n-1, \dots, n-1, n-1).$$

Any constant of $\mathcal{A}$ could be encoded as a 0-1 string representing its binary representation, and these strings, encoding relations and constants of $\mathcal{A}$, could then be concatenated, perhaps, to obtain an encoding of $\mathcal{A}$. Note that there are numerous (an infinite number, in fact) different encoding schemes we might choose to encode the structure $\mathcal{A}$. Consequently, we can now say that some problem is accepted by, for example, some Turing machine if some encoding of it, according to some fixed encoding scheme, is the language accepted by the Turing machine. By "fixed" we do not mean that the same encoding scheme should be used for every problem, just that the same encoding scheme should be used for every structure of our problem. Note also that because a problem is a set of structures closed under isomorphism, there will be a number of strings encoding the "same" structure. (This is the basic approach adopted in complexity theory when abstract decision problems are equated with languages: see [11]).

We are now in a position to identify a complexity class **CC** with a class of problems L. We identify **CC** with L, and write **CC** $= L$ and say that L *captures* **CC**, if, and only if:

- for every problem in L, there is an encoding of this problem in **CC**; and
- for every language in **CC**, there is a problem in L of which this language is the encoding.

We reiterate again that any encoding scheme is only specific to the problem in hand.

Given any language Λ over $\{0,1\}$, we can always obtain a problem Ω of which this language is the encoding; and, moreover, this problem is over the signature $\sigma = \langle M, S \rangle$, where M is a unary relation symbol and S is a binary relation symbol. For any string $\omega \in \Lambda$ of length n, let $\mathcal{A}$ be a σ-structure of size n such that S encodes a successor relation on $\{0, 1, \ldots, n-1\}$, e.g., $\{(0,1),(1,2),(2,3),\ldots,(n-2,n-1)\}$ which yields the ordering $0, 1, 2, \ldots, n-1$, and such that $M(i)$ holds if, and only if, the ith bit of ω, with respect to the ordering given by S, is 1. Note that there are numerous such structures $\mathcal{A}$ corresponding to ω in this way. Let the problem Ω consist of every such σ-structure derived from some string $\omega \in \Lambda$ in this way (and so Ω is closed under isomorphism). As our encoding scheme for Ω, encode any σ-structure $\mathcal{A}$ for which S is a successor relation as the string obtained by reversing the above process, and encode any σ-structure $\mathcal{A}$ for which S is not a successor relation as some fixed string not in Λ (we may assume that there exists such a string). Then, via this encoding scheme, the problem Ω is encoded as the language Λ. We shall return to these "successor" structures presently.

The above discussion suffers from the fact that we have not defined exactly what we mean by an encoding scheme. Rather than becomes embroiled in this issue, let us refer the reader to the discussion in [11] on as to what constitutes a "reasonable" encoding scheme.

We remark that it is usual to say that an input to some Turing machine has size n if the input string consists of n symbols. We always work modulo an encoding scheme and consequently we talk of some structure being input to some Turing machine when strictly we mean that the encoding of this structure is input. Furthermore, if we say that some structure of size n is input to some Turing machine then "size" refers to the size of the structure and not the length of the encoding.

3 Lindström Quantifiers in Descriptive Complexity

We now show how Lindström quantifiers are used in the logical characterization of complexity classes defined by a model of computation not involving oracles. Prompted by these characterizations, we then examine the possibility of using oracle machines to solve problems definable in other related logics formed using Lindström quantifiers.

3.1 In the Absence of Oracles

First-order logic, FO, consists of all formulae formed from atomic formulae (over some signature) using the boolean connectives $\wedge$, $\vee$ and $\neg$, and the quantifiers $\forall$ and $\exists$. We write FO to denote the class of first-order formulae and also the class of problems defined by first-order sentences, and do likewise for any other logic. Whilst it is not difficult to see that any problem in FO is in the complexity class **L**, there are problems in **L** that are not in FO; one such being the problem, over the empty signature, consisting of those structures of even size. Consequently, if we are interested in capturing complexity classes such as **L**, **NL**, **P** and **NP** then first-order logic obviously needs to be extended in some fashion.

One natural extension is to *second-order logic*, SO, which is obtained by allowing the existential and universal quantification of new relation symbols (that is, relation symbols not appearing in the underlying signature). For example, if $\sigma_2 = \langle E \rangle$, where E is a binary relation symbol (and so σ_2-structures can be thought of as undirected graphs via "there is an edge from vertex u to vertex v if, and only if, $E(u,v)$ or $E(v,u)$ holds"; and also as directed graphs via "there is an edge from vertex u to vertex v if, and only if, $E(u,v)$ holds") then the problem

$$3\text{COL} = \{\mathcal{A} \in \text{STRUCT}(\sigma_2) : \text{the undirected graph with vertex set}$$
$$\text{given by } |\mathcal{A}| \text{ and edge set given by the relation } E \text{ can be 3-coloured}\}$$

is defined by the second-order sentence

$$\exists R \exists W \exists B (\forall x (R(x) \vee W(x) \vee B(x))$$
$$\wedge \forall x (\neg (R(x) \wedge W(x)) \wedge \neg (R(x) \wedge B(x)) \wedge \neg (W(x) \wedge B(x)))$$
$$\wedge \forall x \forall y ((E(x,y) \vee E(y,x)) \Rightarrow (\neg (R(x) \wedge R(y)) \wedge \neg (W(x) \wedge W(y))$$
$$\wedge \neg (B(x) \wedge B(y)))))$$

(where R, W and B are new relation symbols af arity 1).

In a seminal result, Fagin [10] proved that a problem is in **NP** if, and only if, it can be defined by a sentence of *existential second-order logic*, Σ_1^1 (that is, the fragment of SO consisting of formulae of the form

$$\exists T_1 \exists T_2 \ldots \exists T_k \phi,$$

where each T_i is a new relation symbol and ϕ is a first-order formula). Stockmeyer [36] extended Fagin's characterization to show that SO captures the Polynomial Hierarchy, **PH**.

Given the facts that FO $\subset$ **L** and that Σ_1^1 is a rather basic fragment of SO, one still has the problem of capturing complexity classes "below" **NP** such as **L**, **NL** and **P**. One method is to restrict the syntax of Σ_1^1 even further, as was done by Grädel [14], but one can also augment FO with appropriate "operators". We illustrate such an approach with the best known of these operators, TC.

Define the signature $\sigma_{2++} = \langle E, C, D \rangle$, where E is a binary relation symbol and C and D are constant symbols, and define the problem TC as

TC = $\{\mathcal{A} \in$ STRUCT(σ_{2++}) : the digraph with vertex set $|\mathcal{A}|$ and

edge set given by the relation E contains a path from vertex C

to vertex $D\}$.

Corresponding to the problem TC is an operator of the same name; that is, an infinite uniform, or vectorized, sequence of *Lindström quantifiers* (whilst we do not define here explicitly what a Lindström quantifier is, we hope that the essence of Lindström quantifiers is gleaned from what follows). The logic $(\pm$TC$)^*$[FO], or *transitive closure logic*, is the closure of FO under the usual first-order connectives and quantifiers, and also the operator TC, with TC applied as follows. Given a formula $\phi(\mathbf{x}, \mathbf{y}) \in (\pm$TC$)^*$[FO], where the variables of the k-tuples $\mathbf{x}$ and $\mathbf{y}$, for some k, are all distinct and free in ϕ, the formula Φ defined as TC$[\lambda \mathbf{xy}\phi](\mathbf{u}, \mathbf{v})$, where $\mathbf{u}$ and $\mathbf{v}$ are k-tuples of (not necessarily distinct) constant symbols and variables, is also a formula of $(\pm$TC$)^*$[FO], with the free variables of Φ being those variables in $\mathbf{u}$ and $\mathbf{v}$, as well as the free variables of ϕ different from those in the tuples $\mathbf{x}$ and $\mathbf{y}$. If Φ is a sentence then it is interpreted in a structure $\mathcal{A} \in$ STRUCT(σ), where σ is the underlying signature, as follows. We build a digraph with vertex set $|\mathcal{A}|^k$ and edge set

$$\{(\mathbf{a}, \mathbf{b}) \in |\mathcal{A}|^k \times |\mathcal{A}|^k : \phi(\mathbf{a}, \mathbf{b}) \text{ holds in } \mathcal{A}\},$$

and say that $\mathcal{A} \models \Phi$ if, and only if, there is a path in this digraph from vertex $\mathbf{u}$ to vertex $\mathbf{v}$ (the semantics can easily be extended to formulae of $(\pm$TC$)^*$[FO]: see, for example, [32] for a more detailed semantic definition). We also denote the fragments of $(\pm$TC$)^*$[FO] where applications of TC do not appear within the scope of a negation sign by TC*[FO], and where at most m applications of TC may be nested by $(\pm$TC$)^m$[FO]: the fragment TC^m[FO] is as expected. We reiterate that TC is essentially an infinite sequence of Lindström quantifiers $\{$TC$_k\}$ where TC$_k$ binds $2k$ free variables in the formula to which it is applied.

In a celebrated result, Immerman [21,22] captured the complexity class **NL** by the logic $(\pm TC)^*[FO]$, but only on the "successor" structures, or to give them their popular name *ordered structures*, encountered earlier. More precisely, he assumed that the logic $(\pm TC)^*[FO]$ has at its disposal a "built-in" binary relation symbol *succ*, different from any symbol of the underlying signature, that is *always* interpreted as a successor relation (i.e., $succ(x, y)$ holds if, and only if, $y = x + 1$) on the domain of any structure, and also two constant symbols, 0 and *max*, that denote the least and greatest elements with respect to the relation *succ*: we denote the resulting logic by $(\pm TC)^*[FO_s]$. Note that sentences of a logic such as $(\pm TC)^*[FO_s]$ might not define problems, i.e., sets of finite structures closed under isomorphism, as, for example, the successor relation might be used "explicitly" in a sentence as it is in the sentence $E(0, max)$ of $FO_s(\sigma_2)$. We simply ignore all such sentences and only ever concern ourselves with sentences defining problems.

However, it is undecidable as to whether a sentence of $(\pm TC)^*[FO_s]$ defines a problem and consequently $(\pm TC)^*[FO_s]$ does not have a recursive syntax and should not really be called a logic (see [18]). Notwithstanding this remark, we continue to call $(\pm TC)^*[FO_s]$ a logic on the grounds that one could regard it as the (bona fide) logic $(\pm TC)^*[FO]$ restricted to the class of ordered structures. But let us press on with our discussion of Immerman's characterization of **NL**.

In fact, Immerman showed that $(\pm TC)^*[FO_s] = TC^1[FO_s] = $ **NL** and that every formula in $(\pm TC)^*[FO_s]$ has a simple normal form; in particular, any problem Ω that is definable by a sentence of $(\pm TC)^*[FO_s]$ can be defined by one of the form

$$TC[\lambda \mathbf{xy} \psi](\mathbf{0}, \mathbf{max}),$$

where: $\mathbf{x}$ and $\mathbf{y}$ are k-tuples of distinct variables, for some k; ψ is a quantifier-free formula of FO_s (actually, a *quantifier-free projection*: see [21]); and $\mathbf{0}$ (resp. $\mathbf{max}$) is the constant symbol 0 (resp. *max*) repeated k times. In such a circumstance we say that there is a *quantifier-free first-order translation with successor* from Ω to TC; and as Ω is an arbitrary problem in **NL** we have that TC is *complete for* **NL** *via quantifier-free first-order translations with successor* (clearly, we may have other translations such as first-order translations with, or without, successor). Such a result sharpens the well-known result that TC is complete for **NL** via log-space reductions [29]. An important point to note is that adopting such a logical viewpoint to complexity theory yields techniques for proving problems complete for some complexity class that were hitherto unavailable (see [34]) and scope for showing that traditional complete problems might not remain complete under restricted logical reductions (see [1]).

One can augment FO (or FO_s) with an operator (or operators) such as TC corresponding to *any* problem (or problems) and examine the class of problems so captured. A variety of such logics have been formed and many well-known complexity classes, including **L**, **NL**, **P** and **NP**, consequently captured (see, for example, the presentation and references in [35]). Of particular interest to us is the logic formed by extending FO_s using an operator corresponding to the problem

$$HP = \{ \mathcal{A} \in \text{STRUCT}(\sigma_{2++}) : \text{ the digraph with vertex set given by}$$
$$|\mathcal{A}| \text{ and edge set given by the relation } E \text{ has a Hamiltonian}$$
$$\text{path from vertex } C \text{ to vertex } D \}.$$

Theorem 1. [5,31] *A problem is in* **NP** *if, and only if, it can be defined by a sentence of the logic* $HP^*[FO_s]$.

In fact, Theorem 1 holds in the absence of the built-in successor relation [5], and there is a normal form result analogous to that for $TC^*[FO_s]$, above [31]. Obviously a more satisfactory result would be that **NP** $= (\pm HP)^*[FO_s]$. However, this would yield as an immediate corollary that **NP** = **co-NP**! Consequently, we are left with the question: *"Can we characterize the class of problems* $(\pm HP)^*[FO_s]$ *in the traditional (Turing-machine-based) setting?"*. Note that such a question does not arise with regard to the analogous logical characterizations of **L**, **NL** and **P** as these complexity classes are closed under complementation.

3.2   Modelling Lindström Quantifiers Using Oracles

Before turning to the question just stated, let us consider the possibilities of modelling applications of Lindström quantifiers using oracles. For concreteness, let us consider a sentence ϕ of the logic $(\pm HP)^*[FO_s]$. There are some points worthy of note, which we return to presently.

(a) There might be a number of applications of HP within ϕ and some of these applications might be nested.
(b) When deciding whether some appropriate structure, of size n, satisfies ϕ, every application of HP involves the building of a digraph of size n^k, for some k.
(c) In constructing ϕ, the first application of the operator HP (in the process of constructing ϕ) is to a first-order formula.

Let us consider a typical standard oracle-machine computation: for concreteness, consider a log-space deterministic oracle Turing machine (DOTM) where the oracle is the problem HP. In a computation by such a Turing machine, oracle strings are (repeatedly) written to the unique write-only oracle tape and when the oracle consultation state is entered, whether the string written on the oracle tape is in the oracle (i.e., whether the string encodes a digraph and two vertices such that there is a Hamiltonian path in the digraph from the first vertex to the second) is reflected in the new state adopted by the machine. The oracle tape is cleared, with the oracle head subsequently positioned over the leftmost cell of the oracle tape; and all this happens in one move of the machine (see, for example, [4] for more on different types of oracle machines). In comparison with the three points remarked upon above, note the following.

(*a*) Whilst there are a number of different oracle access mechanisms in the literature (see, for example, [4]), it is usually the case that oracle calls can not be nested (and this is certainly the case for the standard notion of a log-space DOTM with an **NP** oracle).

(*b*) The lengths of different oracle strings queried in a particular computation might vary dramatically.

(*c*) The resources used to build a specific oracle query in such a computation can be log-space.

So whilst there may be an intuitive link between Lindström quantifiers and oracles, there are a number of points of divergence.

4   Characterizing Hamiltonian Path Logic

As it happens, the points of divergence mentioned in the previous section can often be reconciled. The complexity class $\mathbf{L^{CC}}$ consists of those problems accepted by a standard log-space DOTM with a **CC** oracle.

Theorem 2. [33] *A problem is in* $\mathbf{L^{NP}}$ *if, and only if, it can be defined by a sentence of the logic* $(\pm HP)^*[FO_s]$. *Moreover, any problem in* $\mathbf{L^{NP}}$ *can be defined by a sentence of* $(\pm HP)^*[FO_s]$ *of the form*

$$\exists z_1 \exists z_2 \ldots \exists z_k (HP[\lambda \mathbf{xy}\phi](\mathbf{0}, \mathbf{max}) \wedge \neg HP[\lambda \mathbf{x'y'}\phi'](\mathbf{0}, \mathbf{max}))$$

where ϕ *and* ϕ' *are quantifier-free first-order (actually, quantifier-free projections) and* $z_1, z_2, \ldots, z_k$ *are free in* ϕ *and* ϕ'.

Proof. (*Sketch*) Let Ω be some problem accepted by the log-space DOTM M with an **NP** oracle. Without loss of generality, we may assume that the oracle is HP (essentially because HP is complete for **NP** via 1-*L reductions* [19]). Also, by a result of Wagner [40], we may assume that on any input structure of size n, M makes $O(\log n)$ oracle queries.

Consider a typical computation of M on some input structure of size n. There are some answers (0 or 1) to queries associated with this computation, and as there are $O(\log n)$ queries, we can encode these answers using the binary representations of k values $v_1, v_2, \ldots, v_k \in \{0, 1, \ldots, n-1\}$, for some k. If σ is the underlying signature of Ω, denote by σ' the signature σ with k additional constant symbols $C_1, C_2, \ldots, C_k$.

We shall define a log-space DOTM M' that takes σ'-structures as input and which accepts some σ'-structure $\mathcal{A}'$ if, and only if, M accepts $\mathcal{A} = \mathcal{A}'|_\sigma$ (i.e., the σ-structure obtained from $\mathcal{A}'$ by removing the constants $C_1^{\mathcal{A}'}, C_2^{\mathcal{A}'}, \ldots, C_k^{\mathcal{A}'}$) with the oracle answers as given by the constants $C_1^{\mathcal{A}'}, C_2^{\mathcal{A}'}, \ldots, C_k^{\mathcal{A}'}$.

The DOTM M' proceeds as follows.

Simulate the action of M on $\mathcal{A}$ except that instead of writing on the oracle tape, use the bits of $C_1^{\mathcal{A}'}, C_2^{\mathcal{A}'}, \ldots, C_k^{\mathcal{A}'}$ as the answers to (non-existent) queries.

If the simulation is rejecting then reject else:

  build, and query, a triple (G, a, b), where G is a digraph and a and b
  are distinct vertices of G, with the property that: if
$$(G_1, a_1, b_1), (G_2, a_2, b_2), \ldots, (G_m, a_m, b_m)$$
  are the queries made by M on input $\mathcal{A}$ for which the answer
  (according to $C_1^{\mathcal{A}'}, C_2^{\mathcal{A}'}, \ldots, C_k^{\mathcal{A}'}$) is "yes" then $(G, a, b) \in$ HP if, and
  only if, $(G_1, a_1, b_1), (G_2, a_2, b_2), \ldots, (G_m, a_m, b_m) \in$ HP;
  build, and query, a triple (G', a', b') with the property that: if
$$(G'_1, a'_1, b'_1), (G'_2, a'_2, b'_2), \ldots, (G'_{m'}, a'_{m'}, b'_{m'})$$
  are the queries made by M on input $\mathcal{A}$ for which the answer
  (according to $C_1^{\mathcal{A}'}, C_2^{\mathcal{A}'}, \ldots, C_k^{\mathcal{A}'}$) is "no" then $(G', a', b') \notin$ HP if,
  and only if, $(G'_1, a'_1, b'_1), (G'_2, a'_2, b'_2), \ldots, (G'_{m'}, a'_{m'}, b'_{m'}) \notin$ HP.
  If the answers to these two queries are "yes" and "no",
  respectively, then accept otherwise reject.

The triples (G, a, b) and (G', a', b') can actually be built from $(G_1, a_1, b_1), (G_2, a_2, b_2), \ldots, (G_m, a_m, b_m)$ and $(G'_1, a'_1, b'_1), (G'_2, a'_2, b'_2), \ldots, (G'_{m'}, a'_{m'}, b'_{m'})$ using 1-L reductions [31], and so M' is indeed a log-space DOTM.

We can now encode the computation of M' so that there are two first-order quantifier-free (actually, quantifier-free projections) σ'-formulae $\phi(\mathbf{x}, \mathbf{y})$ and $\phi'(\mathbf{x}', \mathbf{y}')$ with the property that

$$\mathcal{A}' \models \text{HP}[\lambda \mathbf{x}\mathbf{y}\phi](\mathbf{0}, \mathbf{max}) \text{ if, and only if, there is a Hamiltonian path in}$$
$$\text{the digraph } G \text{ from } a \text{ to } b$$

and

$$\mathcal{A}' \models \text{HP}[\lambda \mathbf{x}'\mathbf{y}'\phi'](\mathbf{0}, \mathbf{max}) \text{ if, and only if, there is a Hamiltonian path in}$$
$$\text{the digraph } G' \text{ from } a' \text{ to } b'.$$

Consequently, Ω can be defined by a sentence of the form

$$\exists z_1 \exists z_2 \ldots \exists z_k (\text{HP}[\lambda \mathbf{x}\mathbf{y}\phi](\mathbf{0}, \mathbf{max}) \wedge \neg\text{HP}[\lambda \mathbf{x}'\mathbf{y}'\phi'](\mathbf{0}, \mathbf{max})).$$

Our only remark with regard to the encoding of the computation of M is that $\phi(\mathbf{x}, \mathbf{y})$ and $\phi'(\mathbf{x}', \mathbf{y}')$ do not describe the digraphs G and G' directly. We actually describe these digraphs using formulae of $(\pm\text{DTC})^*[\text{FO}_s]$ where the problem DTC is defined as

$$\text{DTC} = \{\mathcal{A} \in \text{STRUCT}(\sigma_{2++}) : \text{the digraph with vertex set } |\mathcal{A}| \text{ and}$$
$$\text{edge set given by the relation } E \text{ contains a deterministic path}$$
$$\text{from vertex } C \text{ to vertex } D\},$$

with a *deterministic path* between two vertices in a digraph being one where every vertex on the path, except perhaps the last, has out-degree 1 in the digraph (Immerman captured **L** by the logic $(\pm\text{DTC})^*[\text{FO}_s]$ [21]). We then apply the operator HP to these formulae and manipulate the resulting formulae to get our normal form.

The fact that any problem definable in $(\pm\text{HP})^*[\text{FO}_s]$ can be accepted by a log-space DOTM with an **NP** oracle is relatively straightforward to prove.   □

The proof of Theorem 2 yields a number of corollaries regarding the relationship between fragments of $(\pm\text{HP})^*[\text{FO}_s]$ and complexity classes defined by bounding the number of queries made by log-space and polynomial-time oracle machines to an **NP** oracle. We mention one of these corollaries below as an illustration (for others see [32,33]).

The Boolean Hierarchy, **BH**, is defined as follows.

$\mathbf{NP}(0) = \mathbf{P}$;
$\mathbf{NP}(1) = \mathbf{NP}$;
$\mathbf{NP}(2i) = \{X \cap Y : X \in \mathbf{NP}(2i - 1), Y \in \mathbf{co\text{-}NP}\}$;
$\mathbf{NP}(2i + 1) = \{X \cup Y : X \in \mathbf{NP}(2i), Y \in \mathbf{NP}\}$,

for $i \geq 0$, and $\mathbf{BH} = \cup\{\mathbf{NP}(i) : i = 0, 1, \dots\}$. For any function $f(n)$, $\mathbf{L^{NP}}[O(f(n))]$ consists of those problems accepted by a log-space DOTM which makes $O(f(n))$ queries to its **NP** oracle on every input of size n. For any logic L, the logic $Bool(L)$ consists of the closure of L under $\wedge$, $\vee$ and $\neg$. Theorem 2, a simple modification of its proof and a result in [40] (stating that if $\mathbf{L^{NP}}[O(1)] = \mathbf{L^{NP}}$ then the Polynomial Hierarchy collapses to $\mathbf{L^{NP^{NP}}}$) yield the following corollary.

Corollary 1. [33] $Bool(HP^*[FO_s]) = \mathbf{L^{NP}}[O(1)]$; and so $\mathbf{BH} = \mathbf{L^{NP}}[O(1)]$. If $Bool(HP^*[FO_s]) = (\pm HP)^*[FO_s]$ then the Polynomial Hierarchy collapses to $\mathbf{L^{NP^{NP}}}$.   □

The fact that $\mathbf{BH} = \mathbf{L^{NP}}[O(1)]$ was proved in [23].

Write $\mathbf{L^{NP}_{||}}[O(f(n))]$ to denote the complexity class consisting of all those problems accepted by log-space DOTMs which make, on an input of size n, $O(f(n))$ queries to an **NP** oracle such that all queries are made "in parallel", i.e., intuitively, *all* queries are computed before any is made. As established in [40], if either $\mathbf{L^{NP}} = \mathbf{L^{NP}_{||}}[O(\log n)]$ or $\mathbf{L^{NP}_{||}}[O(\log n)] = \mathbf{L^{NP}}[O(1)]$ then the Polynomial Hierarchy collapses. Thus, adhering to the current beliefs in complexity theory, it is unlikely that either of these two events happens. This prompts the question as to whether there is a logical characterization of the complexity class $\mathbf{L^{NP}_{||}}[O(\log n)]$ as a fragment of $(\pm\text{HP})^*[\text{FO}_s]$ (neither $(\pm\text{HP})^*[\text{FO}_s]$ nor $Bool(\text{HP}^*[\text{FO}_s])$ seem likely to suffice).

5   Characterizations of Other Log-Space Oracle Classes

Whilst Theorem 2 resolves the question posed earlier regarding identifying the logic $(\pm\text{HP})^*[\text{FO}_s]$ with a traditional complexity class, it is rather specific in that it deals only with the operator HP; and there is nothing particularly special about HP beyond the fact that it was involved in the first characterization of

NP by what one might call a "Lindström logic". However, from Theorem 2, an analogous result holds for any operator for which the corresponding problem is complete for **NP** via quantifier-free first-order translations with successor (such as 3COL: see [30]).

Such an observation led Gottlob to consider the more general question: *"If Ω is complete for some complexity class **CC** via first-order translations with successor, under what circumstances can we deduce that $(\pm\Omega)^*[FO_s] = \mathbf{L}^{\mathbf{CC}}$?"*, and in [15], Gottlob provided a precise answer which revolves around the concept of smoothness; where a complexity class **CC** is *smooth* if, and only if, $\mathbf{L}^{\mathbf{CC}}(\mathbf{CC}) = \mathbf{L}^{\mathbf{CC}}$, i.e., the class of problems that are $\mathbf{L}^{\mathbf{CC}}$ many-one reducible to a problem in **CC** coincides with $\mathbf{L}^{\mathbf{CC}}$.

Theorem 3. [15] *Let **CC** be some complexity class that is closed under log-space reductions, and let Ω be some problem that is complete for **CC** via first-order translations with successor. Then $(\pm\Omega)^*[FO_s] = \mathbf{L}^{\mathbf{CC}}$ if, and only if, **CC** is smooth.*

Proof. (*Sketch*) For simplicity, assume that Ω is over the signature σ_2 (the general case is similar). We begin by showing that if **CC** is smooth then any problem Γ in $(\pm\Omega)^*[FO_s]$ is in $\mathbf{L}^{\mathbf{CC}}$. We do this by induction on the length of the sentence defining Γ. The only non-trivial case is when this sentence Φ is of the form

$$\Omega[\lambda \mathbf{xy}\phi],$$

where $|\mathbf{x}| = |\mathbf{y}| = k$, for some k, and where the predicate defined by $\phi(\mathbf{x}, \mathbf{y})$ can be checked in $\mathbf{L}^{\mathbf{CC}}$.

Let $\mathcal{A}$ be some $\sigma(\Gamma)$-structure of size n. In order to ascertain whether $\mathcal{A} \models \Phi$, we "build" a σ_2-structure $\mathcal{B}$ whose universe is $|\mathcal{A}|^k$ and whose "binary relation" on $|\mathcal{A}|^k$ contains the pair $(\mathbf{u}, \mathbf{v})$ if, and only if, $\phi(\mathbf{u}, \mathbf{v})$ holds in $\mathcal{A}$. We then check to see whether this structure $\mathcal{B}$ is such that $\mathcal{B} \in \Omega$. Thus, the problem Γ is $\mathbf{L}^{\mathbf{CC}}$ many-one reducible to a problem in **CC**, i.e., $\Gamma \in \mathbf{L}^{\mathbf{CC}}$ because **CC** is smooth.

Now we show that if **CC** is smooth then any problem in $\mathbf{L}^{\mathbf{CC}}$ can be defined by a sentence of $(\pm\Omega)^*[FO_s]$. In the proof of Theorem 2, we used specific properties of the problem HP with regard to 1-L reductions in order to replace a number of oracle queries with just two oracle queries. However, we are now working in a much more general context and, consequently, such problem-specific properties are not at our disposal.

Let M be some log-space DOTM with a **CC** oracle that accepts the problem Γ. Consider a computation of M on the $\sigma(\Gamma)$-structure $\mathcal{A}$ of size n. We may clearly assume that no instantaneous description (ID) occurring in a computation of M is ever repeated in this computation (for us, an ID of M consists of the state, the contents of the work-tape, the work-head position and the input-head position at any particular instant of a computation of M) and that whenever M terminates (which it does on every input), the oracle tape is empty. Note that because M is a log-space DOTM we can:

- encode an ID of M as k values $v_1, v_2, \ldots, v_k \in \{0, 1, \ldots, n-1\}$, for some k;
- "decode" a given k-tuple $\mathbf{u} \in \{0, 1, \ldots, n-1\}^k$ so as to decide in log-space whether it really encodes a legitimate ID of M on some input structure of size n.

Henceforth, we assume that all k-tuples over $\{0, 1, \ldots, n-1\}$ encode potential IDs in some computation of M on an input structure of size n; as, by above, this property can be checked in log-space.

Let $\mathbf{u}$ and $\mathbf{v}$ be IDs of M on input $\mathcal{A}$ such that when M is started in ID $\mathbf{u}$, the next ID it reaches where the oracle tape is empty is $\mathbf{v}$. Then we say that $\mathbf{v}$ is the *direct successor* of $\mathbf{u}$. Additionally, if there is no oracle computation between $\mathbf{u}$ and $\mathbf{v}$, we say that $\mathbf{v}$ is the *strict successor* of $\mathbf{u}$.

Consider the following two problems.

Problem A Given two IDs $\mathbf{u}$ and $\mathbf{v}$ of M and the input structure $\mathcal{A}$, $\mathbf{u}$ and $\mathbf{v}$ are such that either: $\mathbf{v}$ is the strict successor of $\mathbf{u}$; or $\mathbf{v}$ is the direct successor of $\mathbf{u}$, there is an oracle query between $\mathbf{u}$ and $\mathbf{v}$ and the answer to this oracle query is "yes".

Problem B Given two IDs $\mathbf{u}$ and $\mathbf{v}$ of M and the input structure $\mathcal{A}$, $\mathbf{v}$ is the direct successor of $\mathbf{u}$, there is an oracle query between $\mathbf{u}$ and $\mathbf{v}$ and the answer to this oracle query is "no".

Then Problem A is log-space reducible to a problem in **CC**, and Problem B is log-space reducible to a problem in **co-CC**. Thus, by hypothesis, Problems A and B can be defined by $\sigma(\Gamma)$-formulae of $(\pm\Omega)^*[\mathrm{FO}_s]$ of the form

$$\Omega[\lambda\mathbf{xy}\phi(\mathbf{x}, \mathbf{y}, \mathbf{z}, \mathbf{w})] \text{ and } \neg\Omega[\lambda\mathbf{x'y'}\phi'(\mathbf{x'}, \mathbf{y'}, \mathbf{z}, \mathbf{w})],$$

respectively, where ϕ and ϕ' are formulae of FO_s (the tuples of free variables $\mathbf{z}$ and $\mathbf{w}$ encode the IDs). Consequently, there is a $\sigma(\Gamma)$-formula $\Phi(\mathbf{z}, \mathbf{w})$ of $(\pm\Omega)^*[\mathrm{FO}_s]$ of the form

$$\Omega[\lambda\mathbf{xy}\phi] \wedge \neg\Omega[\lambda\mathbf{x'y'}\phi']$$

which holds in some structure $\mathcal{A}$ for some IDs $\mathbf{u}$ and $\mathbf{v}$ if, and only if, $\mathbf{v}$ is the direct successor of $\mathbf{u}$ in the computation of M on input $\mathcal{A}$.

We can now use the operator DTC and our formula $\Phi(\mathbf{z}, \mathbf{w})$ to define the problem consisting of all those $\sigma(\Gamma)$-structures $\mathcal{A}$ such that on input $\mathcal{A}$, there exists an accepting ID that can be reached from the initial ID in the digraph defined by $\Phi(\mathbf{z}, \mathbf{w})$; in other words, we can define the problem Γ. However, as **L** $\subseteq$ **CC**, we can eliminate any applications of the operator DTC at the expense of introducing applications of the operator Ω. Hence, Γ can be defined by a sentence of the logic $(\pm\Omega)^*[\mathrm{FO}_s]$ as required (and the defining sentence just constructed is actually in $(\pm\Omega)^2[\mathrm{FO}_s]$).

In order to prove our theorem, we need to show that if $(\pm\Omega)^*[\mathrm{FO}_s] = \mathbf{L}^{\mathbf{CC}}$ then **CC** is smooth. Suppose that $(\pm\Omega)^*[\mathrm{FO}_s] = \mathbf{L}^{\mathbf{CC}}$. Let the problem Γ be $\mathbf{L}^{\mathbf{CC}}$ many-one reducible to the problem $\Theta \in \mathbf{CC}$. By hypothesis, this $\mathbf{L}^{\mathbf{CC}}$ many-one reduction can be defined by a formula of $(\pm\Omega)^*[\mathrm{FO}_s]$, and the problem

Θ can be defined by a sentence of $(\pm\Omega)^*[\text{FO}_s]$. Hence Γ can be defined by a sentence of $(\pm\Omega)^*[\text{FO}_s]$, and so **CC** is smooth. □

Whilst Theorem 3 gives an answer to the question posed at the beginning of this section, it also provokes the question: *"How can we tell whether a complexity class is smooth or not?"*. Indeed, for all we know *all* complexity classes might be smooth. The most transparently smooth complexity class is **L** as it is well known that the composition of two log-space machines can be replaced with a log-space machine. Essentially, whenever the second machine (in the concatenation) needs some input bit, it computes the input bit required, using log-space, and proceeds accordingly. The important point to note is that it need not "remember" the whole of the output from the first machine (of the concatenation), which might be of polynomial size.

If we were to naively attempt to replicate this construction with the first machine, M_1, a $\mathbf{L}^{\mathbf{CC}}$ machine and the second machine, M_2, a **CC** machine, so as to replace the two machines with one $\mathbf{L}^{\mathbf{CC}}$ machine, M, then (intuitively) we run into problems. Suppose that M is in the process of building some oracle string to query, but half way through this building process it discovers that it needs to compute some input bit to its simulation of M_2, i.e., an output bit from M_1. Naively, it would perform a simulation of M_1; but this might not be possible as M_1 might wish to query some oracle strings in this computation, and the oracle tape of M is blocked with a half-built query already! In fact, there *do* exist complexity classes which are closed under log-space reductions and which are not smooth although we do not prove this fact here (but simply refer the reader to [15]).

Having satisfied ourselves that not all (reasonable) complexity classes are smooth, let us look at some complexity classes which are smooth; and, in particular, at some criteria for smoothness. Many complexity classes (especially deterministic ones) can easily shown to be smooth simply by working through the definition of smoothness with the particular complexity class: such classes include **L**, **POLYLOGSPACE** and **P**. Other complexity classes can easily be shown to be smooth by applying the following lemma; but first two definitions. Reverting back to the traditional definition of a complexity class as a class of languages, we say that a complexity class **CC** is closed under *marked union* if for every two languages Λ_1 and Λ_2 in **CC**, the marked union of Λ_1 and Λ_2, namely the language obtained by inserting a 0 before every string of Λ_1 and a 1 before every string of Λ_2, and then taking the union of the resulting two languages, is also in **CC**. The complexity class $\mathbf{P}^{\mathbf{CC}}$ consists of those problems accepted by a standard polynomial-time DOTM with a **CC** oracle.

Lemma 1. [15] *If* $\mathbf{P}^{\mathbf{CC}} = \mathbf{CC}$ *and* **CC** *is closed under marked union then* **CC** *is smooth.* □

The complexity classes Δ_i^p, for $i \geq 1$, of the Polynomial Hierarchy, and the complexity classes **PSPACE** and k-**EXPTIME**, for $k \geq 1$, can be shown to be smooth by applying Lemma 1.

Other sufficient criteria for smoothness exist. A complexity class is *closed under conjunctions* if, and only if, for any finite set of instances of some problem $\Gamma \in \mathbf{CC}$, the problem of deciding whether every one of these instances is in Γ is also in $\mathbf{CC}$. A problem Ω is $\mathbf{NP}$-*reducible* to some problem Γ if, and only if, there is a non-deterministic polynomial-time transducer M such that on input some $\sigma(\Omega)$-structure $\mathcal{A}$, there is at least one computation of M on $\mathcal{A}$ which outputs a $\sigma(\Gamma)$-structure $\mathcal{B}$ such that $\mathcal{B} \in \Gamma$.

Lemma 2. [15] *If the complexity class* $\mathbf{CC}$ *is closed under* $\mathbf{NP}$-*reductions, under conjunctions and under marked union then* $\mathbf{CC}$ *is smooth.*   □

The complexity classes Σ_i^p and Π_i^p, for $i \geq 1$, of the Polynomial Hierarchy, and the complexity classes $\mathbf{NEXPTIME}$ and $\mathbf{co\text{-}NEXPTIME}$ can be shown to be smooth by applying Lemma 2.

The criteria for smoothness above, whilst widely applicable, do not seem to be strong enough to show that every smooth complexity class is in fact smooth. The most notable example is the complexity class $\mathbf{NL}$ for which smoothness follows from results due to Immerman [21], and for which neither of Lemmas 1 nor 2 seem to be of any use. It would be interesting to derive other criteria from which the smoothness of (complexity classes such as) $\mathbf{NL}$ could be deduced.

Let us end this section by remarking that the proof of Theorem 2 also yielded a normal form result, whereas no analogous normal form result has been forthcoming from the proof of Theorem 3. Perusal of the proof of Theorem 2 yields that the completeness of HP via 1-L reductions played a role; and this might imply that normal form results could be specific to particular problems. However, Gottlob has established general criteria for determining when a normal form result, analogous to that in Theorem 2, exists.

Theorem 4. [15] *Let* $\mathbf{CC}$ *be some complexity class that is closed under* $\mathbf{NP}$-*reductions, conjunctions and marked union, and let* Ω *be some problem that is complete for* $\mathbf{CC}$ *via first-order translations with successor. Then every problem in* $\mathbf{L}^{\mathbf{CC}} = (\pm\Omega)^*[FO_s]$ *can be defined by a sentence of the form*

$$\exists z_1 \exists z_2 \ldots \exists z_k (\Phi \wedge \neg \Phi'),$$

where Φ *and* Φ' *are formulae of* $\Omega^1[FO_s]$ *formed by applying the operator* Ω *to appropriate first-order formulae and where* $z_1, z_2, \ldots, z_k$ *are free in* Φ *and* Φ'.□

The proof of Theorem 4 is similar to the proof of Theorem 2 except that it uses the fact that a complexity class closed under conjunctions and $\mathbf{NP}$-reductions is necessarily closed under conjunctions and disjunctions (defined similarly) to obviate the need for the oracle to be complete for some complexity class via 1-L reductions. It would be interesting to investigate whether there is any sort of relationship between a complexity class having a complete problem via 1-L reductions and a complexity class being closed under operations such as those in Theorem 4.

6   In the Absence of Successor

As remarked earlier, the motivating theorem for the thread of research presented in this paper, Theorem 1, holds in the absence of the built-in successor relation. Consequently, it is natural to ask whether the same can be said of the subsequent results; in particular, Theorem 2. Dawar, Gottlob and Hella have given partial answers in this regard, as we now explain; but first we require some definitions.

We denote by:

- L^k the fragment of FO which consists of those formulae whose variables, both bound and free, are among $x_1, x_2, \ldots, x_k$;
- $L^k_{\infty\omega}$ the closure of L^k under the operations of conjunction and disjunction applied to arbitrary (finite and infinite) sets of formulae (with obvious semantics);
- $L^\omega_{\infty\omega}$ the union $\bigcup_{k=0}^{\infty} L^k_{\infty\omega}$.

The logic $L^\omega_{\infty\omega}$ was introduced by Barwise [3] and is known as *bounded-variable infinitary logic*. It plays an important role in finite model theory (see, for example, [9]).

However, we introduce $L^\omega_{\infty\omega}$ here for a specific reason. Bounded-variable infinitary logic can be extended by a set of Lindström quantifiers Ω (possibly infinite and not necessarily in the form of uniform sequences corresponding to problems), just as FO was (and with similar semantics), to yield the logic $(\pm\Omega)^*[L^\omega_{\infty\omega}]$. Dawar and Hella [8] have established certain properties of logics of the form $(\pm\Omega)^*[L^\omega_{\infty\omega}]$, or more specifically, of fragments of such logics where the number of quantifiers of Ω appearing in any formula is finite, which are of direct relevance to the investigations presented here.

Let $\mathcal{A}$ be some structure and let $\mathbf{u} = (u_1, u_2, \ldots, u_k) \in |\mathcal{A}|^k$. The *basic equality type* of $\mathbf{u}$ is the formula

$$\bigwedge_{(i,j)\in S} (x_i = x_j) \wedge \bigwedge_{(i,j)\in T} (\neg(x_i = x_j)),$$

where $S = \{(i,j) : i < j \text{ and } u_i = u_j\}$ and $T = \{(i,j) : i < j \text{ and } u_i \neq u_j\}$.

Let ϕ be a formula of $(\pm\Omega)^*[L^\omega_{\infty\omega}]$, where Ω is a set of Lindström quantifiers. Then ϕ is a *basic flat formula* if it is atomic or if it is formed by applying a quantifier $Q \in \Omega$ to some first-order quantifier-free formulae. Also, ϕ is in *flat normal form* if it is obtained from basic flat formulae by successive applications of the connectives $\vee$, $\wedge$ and $\neg$ and the first-order quantifiers $\forall$ and $\exists$. For example, the formula $\mathrm{HP}[\lambda\mathbf{xy}\phi](\mathbf{0}, \mathbf{max})$ of Theorem 2 is a basic flat formula, whereas the formula $\exists z_1 \exists z_2 \ldots \exists z_k (\mathrm{HP}[\lambda\mathbf{xy}\phi](\mathbf{0}, \mathbf{max}) \wedge \neg\mathrm{HP}[\lambda\mathbf{x}'\mathbf{y}'\phi'](\mathbf{0}, \mathbf{max}))$ of Theorem 2 is in flat normal form.

We are now in a position to state Dawar and Hella's result (which although stated in [7] is proved in [8]).

Theorem 5. [7,8] *Let σ_ϵ be the empty signature, and denote the unique structure of size n over σ_ϵ by $\mathcal{A}_n$. Let Ω_0 be a finite set of Lindström quantifiers.*

(a) *For every n, there exists a sentence $\eta_n \in (\pm\Omega_0)^*[L^k]$ such that for every n', $\mathcal{A}_{n'} \models \eta_n$ if, and only if, $\mathcal{A}_n$ and $\mathcal{A}_{n'}$ satisfy the same sentences of $(\pm\Omega_0)^*[L^k_{\infty\omega}]$.*

(b) *The sentence η_n is of the form*

$$\bigwedge_{1 \leq i \leq m} \exists x_1 \ldots \exists x_k \psi_i \wedge \forall x_1 \ldots \forall x_k \bigvee_{1 \leq i \leq m} \psi_i \wedge \bigwedge_{1 \leq j \leq r} \forall x_1 \ldots \forall x_k (\phi_j \leftrightarrow \gamma_j),$$

where $\psi_1, \psi_2, \ldots, \psi_m$ are the basic equality types of k-tuples of $\mathcal{A}_n$, each formula γ_j is a disjunction of some basic equality types and each formula ϕ_j is a basic flat formula.

(c) *Given some $\mathbf{u} \in |\mathcal{A}_n|^l$, there is a formula $\eta_{n,\mathbf{u}}$ of the form*

$$\eta_n \wedge \forall x_{l+1} \ldots \forall x_k \gamma,$$

where γ is a disjunction of some basic equality types, such that for every $\mathcal{A}_{n'}$ and $\mathbf{u}' \in |\mathcal{A}_{n'}|^l$,

$$\mathcal{A}_{n'} \models \eta_{n,\mathbf{u}}(\mathbf{u}') \text{ if, and only if, the expansions } \langle \mathcal{A}_n, \mathbf{u} \rangle \text{ and } \langle \mathcal{A}_{n'}, \mathbf{u}' \rangle$$
$$\text{satisfy exactly the same sentences of } (\pm\Omega_0)^*[L^k_{\infty\omega}].$$

□

Corollary 2. [7] *Let Ω_0 be a finite set of Lindström quantifiers and let σ_ϵ be the empty signature. For every formula $\phi(x_1, x_2, \ldots, x_l) \in (\pm\Omega_0)^*[L^\omega_{\infty\omega}(\sigma_\epsilon)]$, there exists a formula $\psi(x_1, x_2, \ldots, x_l) \in (\pm\Omega_0)^1[FO(\sigma_\epsilon)]$ such that ϕ and ψ are equivalent.*

Proof. There is clearly some k such that $\phi \in (\pm\Omega_0)^*[L^k_{\infty\omega}(\sigma_\epsilon)]$. Fix n and let $\psi_1, \psi_2, \ldots, \psi_m$ be the basic equality types of k-tuples of elements of $\mathcal{A}_n$. Let η_n be the sentence as in the statement of Theorem 5, and for every $\mathbf{u} \in |\mathcal{A}_n|^l$, let $\eta_{n,\mathbf{u}}$ be as in Theorem 5. Define $F = \{(n, \mathbf{u}) : \mathbf{u} \in |\mathcal{A}_n|^l \text{ such that } \mathcal{A}_n \models \phi(\mathbf{u})\}$ and define ψ as

$$\bigvee_{(n,\mathbf{u}) \in F} \eta_{n,\mathbf{u}}.$$

Consider any n and $\mathbf{u} \in |\mathcal{A}_n|^l$. Suppose that $\mathcal{A}_n \models \phi(\mathbf{u})$. So, $(n, \mathbf{u}) \in F$, $\eta_{n,\mathbf{u}}$ is a disjunct of ψ and $\mathcal{A}_n \models \eta_{n,\mathbf{u}}(\mathbf{u})$. Thus, $\mathcal{A}_n \models \psi(\mathbf{u})$. Conversely, suppose that $\mathcal{A}_n \models \psi(\mathbf{u})$. So, $\mathcal{A}_n \models \eta_{n',\mathbf{u}'}(\mathbf{u})$, for some $(n', \mathbf{u}') \in F$, with the result that $\langle \mathcal{A}_n, \mathbf{u} \rangle$ and $\langle \mathcal{A}_{n'}, \mathbf{u}' \rangle$ satisfy the same sentences of $(\pm\Omega_0)^*[L^k_{\infty\omega}]$. In particular, $\mathcal{A}_n \models \phi(\mathbf{u})$. The result follows. □

The following is immediate from Corollary 2.

Corollary 3. [7] *If Ω is a set of Lindström quantifiers such that every quantifier can be evaluated in **NP** then every sentence of $(\pm\Omega)^*[FO(\sigma_\epsilon)]$ is equivalent to a Boolean combination of **NP** properties.* □

Given a language Λ over $\{0, 1\}$,

$$tally(\Lambda) = \{1^n : \text{the binary representation of } n \in 1\Lambda\}.$$

The Linear Exponential Boolean Hierarchy, **EBH**, is defined as follows.

EBH$(0) = $ **DETIME** $= $ **DTIME**$(2^{O(n)})$;
EBH$(1) = $ **NETIME** $= $ **NTIME**$(2^{O(n)})$;
EBH$(2i) = \{X \cap Y : X \in $ **EBH**$(2i - 1), Y \in $ **co-NETIME**$\}$;
EBH$(2i + 1) = \{X \cup Y : X \in $ **EBH**$(2i), Y \in $ **NETIME**$\}$,

for $i \geq 0$, and **EBH** $= \cup\{$**EBH**$(i) : i = 0, 1, \ldots\}$. The Full Exponential Boolean Hierarchy, **EXPBH**, is defined as follows.

EXPBH$(0) = $ **DEXPTIME** $= $ **DTIME**$(2^{O(poly(n))})$;
EXPBH$(1) = $ **NEXPTIME** $= $ **NTIME**$(2^{O(poly(n))})$;
EXPBH$(2i) = \{X \cap Y : X \in $ **EXPBH**$(2i - 1), Y \in $ **co-NEXPTIME**$\}$;
EXPBH$(2i + 1) = \{X \cup Y : X \in $ **EXPBH**$(2i), Y \in $ **NEXPTIME**$\}$,

for $i \geq 0$, and **EXPBH** $= \cup\{$**EXPBH**$(i) : i = 0, 1, \ldots\}$.

The hierarchies **EBH** and **EXPBH** can be regarded as exponential versions of the Boolean Hierarchy. Indeed, we can go further and associate with many complexity classes **CC** contained in the Polynomial Hierarchy, a *linear exponential version* $E(\mathbf{CC})$ and a *full exponential version* $EXP(\mathbf{CC})$. Such an association is detailed in Fig. 1 below, with the motivation behind such an association given in the subsequent lemma. The definitions of the complexity classes have either been given or are obvious, except that $\{\Sigma_i^e, \Pi_i^e : i = 0, 1, \ldots\}$ (resp. $\{\Sigma_i^{exp}, \Pi_i^{exp} : i = 0, 1, \ldots\}$) are the classes of the *Linear* (resp. *Full*) *Exponential Hierarchy* **EH** (resp. **EXPH**), built around **ETIME** and **NETIME** (resp. **EXPTIME** and **NEXPTIME**) just as the Polynomial Hierarchy is built around **P** and **NP**.

Basic complexity class CC	Linear exponential version $E(\mathbf{CC})$	Full exponential version $EXP(\mathbf{CC})$
P	**ETIME**	**EXPTIME**
NP	**NETIME**	**NEXPTIME**
Σ_i^p	Σ_i^e	Σ_i^{exp}
Π_i^p	Π_i^e	Π_i^{exp}
PH	**EH**	**EXPH**
BH(i)	**EBH**(i)	**EXPBH**(i)
BH	**EBH**	**EXPBH**
L	**LINSPACE**	**PSPACE**
NL	**NLINSPACE**	**PSPACE**
L$^{\mathbf{NP}}$	**LINSPACE**$^{\mathbf{NP}}$	**PSPACE**$^{\mathbf{NP}}$

Figure 1. Linear and full exponential versions of complexity classes.

Lemma 3. [16] *For each basic complexity class* **CC** *in the first column of Figure 1:*

- *the closure under polynomial-time many-one reductions of* $E(\textbf{CC})$ *is the complexity class* $EXP(\textbf{CC})$; *and*
- *for every language* $\Lambda \in \textbf{CC}$, $tally(\Lambda) \in \textbf{CC}$ *if, and only if,* $\Lambda \in E(\textbf{CC})$.

□

We are now ready to establish the results of Dawar, Gottlob and Hella.

Theorem 6. [7] *If there exists a family* Ω *of Lindström quantifiers where each quantifier in* Ω *can be evaluated in* **NP** *and where* $(\pm\Omega)^*[FO] = \textbf{L}^{\textbf{NP}}$ *then:*

- **EBH** = **LINSPACE$^{\textbf{NP}}$** *and* **EBH** *collapses; and*
- **EXPBH** = **PSPACE$^{\textbf{NP}}$** *and* **EXPBH** *collapses.*

Proof. Let Λ be a language in **LINSPACE$^{\textbf{NP}}$**. By Lemma 3, $tally(\Lambda) \in \textbf{L}^{\textbf{NP}}$. Define the problem Γ over σ_ϵ to be such that $tally(\Lambda)$ is the natural encoding of Γ. By hypothesis, $\Gamma \in (\pm\Omega)^*[\text{FO}] = \textbf{L}^{\textbf{NP}}$; and so by Corollary 3,

$$tally(\Lambda) \in \textbf{EBH}_k, \text{ for some } k.$$

But **LINSPACE$^{\textbf{NP}}$** has complete problems, and consequently **LINSPACE$^{\textbf{NP}}$** $\subseteq$ **EBH$_k$**. As **EBH** $\subseteq$ **LINSPACE$^{\textbf{NP}}$**, the first part of the theorem follows.

However, the second part of the theorem also follows as, by Lemma 3, the complexity classes **EXPBH$_k$**, **EXPBH** and **PSPACE$^{\textbf{NP}}$** are the closures of the complexity classes **EBH$_k$**, **EBH** and **LINSPACE$^{\textbf{NP}}$**, respectively.   □

Any collapses of the form detailed in Theorem 6 would cause not inconsiderable surprise amongst many complexity theorists. Therefore, we are led to speculate that the complexity class $\textbf{L}^{\textbf{NP}}$ can not be captured on the class of all structures by a "Lindström logic". It would be extremely interesting if this result could be established. However, a great improvement of Theorem 6 would be if the premise in the theorem yielded a complexity-theoretic collapse "lower down", say of the Polynomial Hierarchy.

We close by noting that things are different for some complexity classes "beyond" $\textbf{P}^{\textbf{NP}}$. Amalgamating results due to Gurevich [17] and Blass and Gurevich [2], Dawar [6] observed that any recursively presented complexity class containing $\textbf{P}^{\textbf{NP}}$ and closed under compositions is recursively indexable, and so there is a logic capturing this class. Other results in [6] then yield that if the complexity class is *bounded* (see [6]), then it can be captured by a "Lindström logic". So, for example, $\textbf{L}^{\Sigma_2^p}$ can be captured by a "Lindström logic" (as it is bounded).

Acknowledgements

I am indebted to Argimiro Arratia-Quesada, Anuj Dawar, Richard Gault and Juha Nurmonen for their comments on a first draft of this paper.

References

1. E. Allender, J. Balcázar and N. Immerman, A first-order isomorphism theorem, *Lecture Notes in Computer Science Vol.* 665 (1993) 163–174.
2. A. Blass and Y. Gurevich, Equivalence relations, invariants and normal forms, *SIAM J. Comput.* **13** (1984) 682–689.
3. J. Barwise, On Moschovakis closure ordinals, *J. Symbolic Logic* **42** (1997) 292–296.
4. J.F. Buss, Alternations and space-bounded computations, *J. Comput. Systems Sci.* **36** (1988) 351–378.
5. E. Dahlhaus, Reduction to NP-complete problems by interpretations, *Lecture Notes in Computer Science Vol.* 171, Springer-Verlag (1984) 357–365.
6. A. Dawar, Generalized quantifiers and logical reducibilities, *J. Logic Computat.* **5** (1995) 213–226.
7. A. Dawar, G. Gottlob and L. Hella, Capturing relativized complexity classes without order, *Mathematical Logic Quarterly*, 44 (1998) 109–122.
8. A. Dawar and L. Hella, The expressive power of finitely many generalized quantifiers, *Inform. and Computat.* **123** (1995) 172–184.
9. H.D. Ebbinghaus and J. Flum, *Finite Model Theory*, Springer-Verlag (1995).
10. R. Fagin, Generalized first-order spectra and polynomial-time recognizable sets, in: *Complexity of Computation* (ed. R.M. Karp), *SIAM-AMS Proceedings* **7** (1974) 43–73.
11. M.R. Garey and D.S. Johnson, *Computers and Intractability: a Guide to the Theory of NP-Completeness*, W.H. Freeman and Co. (1979).
12. Y.V. Glebskii, D.I. Kogan, M.I. Liogonki and V.A. Talanov, Range and degree of realizability of formulas in the restricted predicate calculus, *Cybernetics* **5** (1969) 142–154.
13. E. Grädel, On logical descriptions of some concepts in structural complexity theory, *Lecture Notes in Computer Science Vol.* 440 (1990) 163–175.
14. E. Grädel, Capturing complexity classes by fragments of second-order logic, *Theoret. Comput. Sci.* **101** (1992) 35–57.
15. G. Gottlob, Relativized logspace and generalized quantifiers over finite ordered structures, *J. Symbolic Logic* **62** (1997) 545–574.
16. G. Gottlob, N. Leone and H. Veith, Second order logic and the weak exponential hierarchies, *Lecture Notes in Computer Science Vol.* 969, Springer-Verlag (1995) 66–81.
17. Y. Gurevich, Toward logic tailored for computational complexity, *Lecture Notes in Mathematics Vol.* 1104, Springer-Verlag (1984) 175–216.
18. Y. Gurevich, Logic and the challenge of computer science, in: *Trends in Theoretical Computer Science* (ed. E. Börger), Computer Science Press (1988) 1–57.
19. J. Hartmanis, N. Immerman and S. Mahaney, One-way log-tape reductions, *Proc. 19th Symp. Foundations of Computer Science*, IEEE Press (1978) 65–71.
20. N. Immerman, Relational queries computable in polynomial time, *Inform. Control* **68** (1986) 86–104.
21. N. Immerman, Languages that capture complexity classes, *SIAM J. Comput.* **16** (1987) 760–778.

22. N. Immerman, Nondeterministic space is closed under complementation, *SIAM J. Comput.* **17** (1988) 935–938.
23. J. Köbler, U. Schöning and K.W. Wagner, The difference and the truth-table hierarchies for NP, *RAIRO Inform. Theory* **21** (1987) 419–435.
24. P. Lindström, First order predicate logic with generalized quantifiers, *Theoria* **32** (1966) 186–195.
25. P. Lindström, On extensions of elementary logic, *Theoria* **35** (1969) 1–11.
26. J.A. Makowsky and Y.B. Pnueli, Oracles and quantifiers, *Lecture Notes in Computer Science Vol.* 832 (1994) 189–222.
27. J.A. Makowsky and Y. Pnueli, Logics capturing oracle complexity classes uniformly, *Lecture Notes in Computer Science Vol.* 960 (1995) 463–479.
28. J.A. Makowsky and Y. Pnueli, Computable quantifiers and logics over finite structures, in: *Quantifiers: Logics, Models and Computation, Volume 1* (ed. M. Krynicki et al.), Kluwer (1995) 313–357.
29. W. Savitch, Maze recognizing automata and nondeterministic tape complexity, *J. Comput. System Sci.* **7** (1973) 389–403.
30. I.A. Stewart, Comparing the expressibility of languages formed using NP-complete operators, *J. Logic Computat.* **1** (1991) 305–330.
31. I.A. Stewart, Using the Hamiltonian path operator to capture NP, *J. Comput. System Sci.* **45** (1992) 127–151.
32. I.A. Stewart, Logical characterizations of bounded query classes I: logspace oracle machines, *Fund. Informat.* **18** (1993) 65–92.
33. I.A. Stewart, Logical characterizations of bounded query classes II: polynomial-time oracle machines, *Fund. Informat.* **18** (1993) 93–105.
34. I.A. Stewart, Methods for proving completeness via logical reductions, *Theoret. Comput. Sci.* **118** (1993) 193–229.
35. I.A. Stewart, Logical description of monotone NP problems, *J. Logic Computat.* **4** (1994) 337–357.
36. L. Stockmeyer, The polynomial hierarchy, *Theoret. Comput. Sci.* **3** (1976) 1–22.
37. R. Szelepcsényi, The method of forced enumeration for nondeterministic automata, *Acta Informat.* **26** (1988) 279–284.
38. B.A. Trakhtenbrot, Impossibility of an algorithm for the decision problem on finite classes, *Doklady* **70** (1950) 569–572.
39. M. Vardi, Complexity of relational query languages, *Proc. 14th ACM Ann. Symp. on Theory of Computing* (1982) 137–146.
40. K.W. Wagner, Bounded query classes, *SIAM J. Comput.* **19** (1990) 833–846.

Generalized Quantifiers in Logic Programs[*]

Thomas Eiter[1], Georg Gottlob[2], and Helmut Veith[2,3]

[1] Institut für Informationssysteme, Abteilung Wissensbasierte Systeme
TU Wien, A-1040 Wien, Austria
eiter@kr.tuwien.ac.at
[2] Institut für Informationssysteme, Abteilung Datenbanken und AI
TU Wien, A-1040 Wien, Austria
(gottlob|veith)@dbai.tuwien.ac.at
[3] School of Computer Science, Carnegie Mellon University, Pittsburgh, USA.
veith@cs.cmu.edu

Abstract. Generalized quantifiers are an important concept in model-theoretic logic which has applications in different fields such as linguistics, philosophical logic and computer science. In this paper, we consider a novel application in the field of logic programming, which has been presented recently. The enhancement of logic programs by generalized quantifiers is a convenient tool for interfacing extra-logical functions and provides a natural framework for the definition of modular logic programs. We survey the expressive capability of syntactical classes of logic programs with generalized quantifiers over finite structures, and pay particular attention to modular logic programs. Moreover, we study the complexity of such programs. It appears that modular logic programming has the expressive power of second-order logic and captures the polynomial hierarchy, and different natural syntactical fragments capture the classes therein. The program complexity parallels the expressive power in the weak exponential hierarchy. Modular logic programming proves to be a rich formalism whose expressiveness and complexity can be controlled by efficiently recognizable syntactic restrictions.

1 Introduction

Generalized quantifiers are a well-known concept for enhancing the expressive capabilities of a logical language [44,34,7,59]. They have been introduced since in many contexts, the standard quantifiers "for all individuals" and "for some individual" are not strong enough for a proper description of the state of affairs. Informally, a generalized quantifier $Q\mathbf{x}$ is a collection C of structures for a language $\mathcal{L}$, and $Q\mathbf{x}\varphi(\mathbf{x}, \mathbf{y})$ evaluates on a structure $\mathfrak{A}$ to true for elements $\mathbf{y} \in A$, if the structure $(A, \{\mathbf{x} \in A \mid \mathfrak{A} \models \varphi(\mathbf{x}, \mathbf{y})\})$ belongs to C.

---

[*] This work was partially supported by the Austrian Science Fund Project N Z29-INF, and the Max Kade Foundation. Most of this research has been carried out while the first author was with the University of Gießen, Germany, and the third author was with TU Wien.

J. Väänänen (Ed.): ESSLLI'97, LNCS 1754, pp. 72–98, 1999.

In 1957, Mostowski [44] initiated the field, and stimulated a lot of work on generalized quantifiers during the 60's and early 70's. Another approach to generalized quantifiers from 1959 are Henkin quantifiers, which have been introduced in [27] and subsequently studied in numerous papers, e.g. [18,64,5,9,33,50]. Unifying the two approaches, the most prominent work from this area is the seminal paper of Lindström [34], whose formalization of generalized quantifiers is mostly used to date. After diminished attention for some time, generalized quantifiers gained in the 90's increasing interest in applications, from which new research issues emerge. The applications of generalized quantifiers range over different fields, including linguistics, philosophical logic and computer science.

In linguistics and philosophical logic, it has been recognized that the quantification theory of first-order logic is insufficient as a representation formalism for natural language statements, cf. [28]. Montague [43] made an attempt to properly treat quantifier phrases in English, the idea to which, as pointed out in [29], originally dates back to Russell [48]. Building on the work of Montague [43], Barwise and Cooper have presented an approach which adopts generalized quantifiers in a natural language formalization [6].

In computer science, generalized quantifiers are used in different areas. On the foundational side, they have been considered extensively in descriptive complexity theory as a tool for designing logical languages for capturing computational complexity classes. For example, restricted Henkin quantifiers capture NL [9] over ordered structures, while standard Henkin quantifiers capture Θ_2^p over ordered structures [22]; transitive closure logic captures NL over ordered structures, while deterministic transitive closure logic captures L over ordered structures [30,25]; and, various extensions of first-order logic by quantifiers such as the Hamiltonian Path Quantifier capture NP or Θ_2^p (i.e., logspace with an NP oracle) over ordered structures, [53,54,22]. The relationship of oracles, generalized quantifiers and subprograms was also investigated in [38,37,40,41]. See [31,13] for a comprehensive overview. Notice that all syntactic complexity classes can be captured by generic generalized quantifiers [36,39,61].

In the context of databases, generalized quantifiers have been used as a means for increasing the expressive capability of database query languages. In [1], the transitive closure quantifier was applied ad hoc since connectivity cannot be expressed in Codd's relational calculus. A more systematic treatment is given in [3,47], where a formalization of subqueries in SQL by generalized quantifiers is presented, and occurrences of generalized quantifiers in existing database query languages are discussed. Further applications of generalized quantifiers in computer science are in software engineering, where generalized quantifiers are used in the specifications of the Z and B notation [52,65]. Moreover, generalized quantifiers have also been investigated in the context of fuzzy logic [26], which has become popular in computer science more recently.

In this paper, we address another application of generalized quantifiers in computer science; namely, in the field of logic programming. As pointed out in [16], enhancing logic programming by generalized quantifiers provides an elegant and appealing method for handling the following two problems. Firstly, the problem of interfacing external (non-logical) functions in a logic program, of which no other information than a semantical description is available. Such a situation is present e.g. with library routines, for which usually only a functional specification is known. Secondly, the problem of modular programming, where a formalism for the synthesis of program modules into logic programs is sought.

The two problems are not unrelated; the latter problem can be seen as an instance of the former, in which modules are functions for which information about the implementation in terms of proof theory etc. is available. For both problems, so far mainly ad hoc solutions have been proposed. Modular logic programming has been approached from two main directions [10]. The one direction is to build an algebra of programs, in which composition operators are available for combining two programs into a single one (cf. [19,45]). The other direction resorts to using concepts of modal logic and embedded implications for defining the meaning of subgoals which access external functions (cf. [42]). Both approaches, however, have problems in handling negation, and typically, negation is disregarded for combined programs. This is a severe shortcoming, since negation is an indispensable element of a logic-based programming language to be sufficiently expressive for solving problems in practice.

Our extension of logic programs by generalized quantifiers is a natural enrichment of the formalism which allows for a natural and appealing solution to the problem of incorporating external functions into logic programs. As an application, we obtain a semantics for modular logic programming which is conceptually clean and, as a major advantage, harbors a full treatment of negation in the style of ordinary logic programs with negation.

Roughly, the idea of the approach is to allow in the bodies of rules besides standard logical literals also formulas $Q\mathbf{x}[R(\mathbf{x})]$, which describe the application of a generalized quantifier $Q\mathbf{x}$ to a structure defined by the predicate R. Informally, $Q\mathbf{x}$ can been seen as an interface to an external function whose argument is the predicate R. For any extension of R, the function resp. quantifier $Q\mathbf{x}$ evaluates to either true or false, and thus succeeds or fails as a subgoal in a rule body. An extension to functions with multiple arguments is straightforward.

In the present paper, we survey the expressive capability of logic programs with different classes of generalized quantifiers, where we focus mainly on finite structures. We pay particular attention to various natural syntactic fragments of the formalism. The programs may use generalized quantifiers of a particular complexity, and in particular generalized quantifiers which are definable by logic programs.

It appears that normal logic programs with quantifiers in NP capture the class Π_2^p of the polynomial hierarchy; thus, by adding such quantifiers, the expressive capability of normal programs increases from co-NP [51] by one level in the hierarchy. As a consequence, modular logic programs capture Π_2^p as well. The expressive capability further increases by allowing nested program modules. The natural extension of the formalism to nested modular logic programs captures the polynomial hierarchy, and is therefore as expressive as second-order logic over finite structures. In fact, the equivalence of nested modular logic programs to second order logic holds even for arbitrary structures. Moreover, natural syntactic fragments of the formalism capture the classes Π_k^p, $k \geq 1$, of the polynomial hierarchy, and over ordered finite structures, also the classes Δ_k^p and Θ_k^p, for $k \geq 1$.

The rest of this paper is organized as follows. In the next section, we recall some basic concepts and fix notation. In Section 3, we consider logic programs with generalized quantifiers, and in Section 4, we describe a formalism for modular logic programming based on logic programs with generalized quantifiers. Thereafter, we address the expressive capability of the formalism in Section 5 and the computational complexity in Section 6. Programs with nested modules are considered in Section 7. Finally, Section 8 concludes the paper.

2   Preliminaries and Notation

Notation. Letters $P, Q, \ldots$ denote predicates, lower case letters x, y, z variables, $a, b, c, \ldots$ constants and $f, g \ldots$ functions. The bold face version $\mathbf{P}$ of a predicate symbol P denotes a list $P_1, \ldots, P_m$ of predicate symbols, and similar for variable, function, and constant symbols. Fraktur letters $\mathfrak{A}, \mathfrak{B}, \ldots$ denote logical structures. Sets of structures are denoted by capital Greek letters Π, Σ or C, and classes or mappings thereof by $\mathcal{M}, \mathcal{Q}, \ldots$; lower case Greek letters $\tau, \sigma, \ldots$ denote signatures.

Logic. A signature[1] is a sequence $\tau = (P_1^{a_1}, \ldots, P_k^{a_k}, f_1^{b_1}, \ldots, f_l^{b_l}, c_1, \ldots, c_m)$ where the P_i are relational symbols of arity $a_i \geq 0$, the f_i are functions with $b_i \geq 1$ arguments, and the c_i are constants. τ is *relational*, if it contains only relational symbols.

A *structure* $\mathfrak{A}$ over τ is denoted by $(A, P_1^{\mathfrak{A}}, \ldots, P_k^{\mathfrak{A}}, f_1^{\mathfrak{A}}, \ldots, f_l^{\mathfrak{A}}, c_1^{\mathfrak{A}}, \ldots, c_m^{\mathfrak{A}})$. A is called the *universe* or *domain* of $\mathfrak{A}$, and denoted $|\mathfrak{A}|$. $\mathfrak{A}$ is finite if $|\mathfrak{A}|$ is finite. The set of all structures over τ is denoted by $\mathrm{Struct}(\tau)$.

Let $\mathfrak{A}, \mathfrak{B} \in \mathrm{Struct}(\tau)$, s.t. $|\mathfrak{A}| = |\mathfrak{B}|$. Then $\mathfrak{A} \subseteq \mathfrak{B}$, if $P_i^{\mathfrak{A}} \subseteq P_i^{\mathfrak{B}}$ for $1 \leq i \leq k$, $f_i^{\mathfrak{A}} = f_i^{\mathfrak{B}}$ for $1 \leq i \leq l$, and $c_i^{\mathfrak{A}} = c_i^{\mathfrak{B}}$ for $1 \leq i \leq m$. A *computational problem* over signature τ is a set $\Pi \subseteq \mathrm{Struct}(\tau)$, s.t. Π is closed under isomorphisms. Π is *monotone*, if $\mathfrak{A} \subseteq \mathfrak{B}$ and $\mathfrak{A} \in \Pi$ implies $\mathfrak{B} \in \Pi$.

For a relational signature τ and integer l, $\tau^{(l)} = (P_1^{la_1}, \ldots, P_k^{la_k})$ is called the $l-ary$ *vectorization* of τ. For a computational problem Π, let $\Pi^{(l)}$ denote the problem Π over $l-$tuples, i.e. over universe $\underbrace{n \times \cdots \times n}_{l \text{ times}}$.[2] Let $\mathfrak{A}$ be a relational structure, and $U \subseteq |\mathfrak{A}|$. Then the restriction of $\mathfrak{A}$ to U, in symbols $\mathfrak{A}|U$, is the structure $(U, P_1^{\mathfrak{A}} \cap U^{a_1}, \ldots, P_k^{\mathfrak{A}} \cap U^{a_k})$. For a τ-structure $\mathfrak{A}$ and a signature τ_0

---

[1] In contrast to the usual convention in finite model theory, we do not disregard function symbols in our language at this stage, because they are commonly available in logic programming.

[2] This is a coherent definition, since we deal only with problems which are closed under isomorphisms.

contained in τ, $\mathfrak{A}|\tau_0$ is the τ_0-structure obtained from $\mathfrak{A}$ by removing all relations, functions, and constants not contained in τ_0.

The set of all finite models of a formula Ψ is denoted by $\mathrm{Mod}(\Psi)$.

Let $\varphi(x_1, \ldots, x_n)$ be a formula with free variables $x_1, \ldots, x_n$, and let $\mathfrak{A}$ be a structure. Then $\varphi^{\mathfrak{A}}$ denotes the $n-$ary relation $\{(d_1, \ldots, d_n) \mid \mathfrak{A} \models \varphi(d_1, \ldots, d_n)\}$.

Let $\mathcal{L}$ be a syntactic fragment of first-order logic. Given signatures τ, σ and a natural number k, a k-ary interpretation I of τ into σ is a definition of the $\sigma^{(k)}$ relations in terms of τ, i.e., a tuple of $\mathcal{L}$ formulas, s.t. for each predicate symbol R in σ with arity r, I contains a formula φ_R over τ with rk free variables which defines R^{rk}. For a structure $\mathfrak{A} \in \mathrm{Struct}(\tau)$, $I(\mathfrak{A})$ denotes the structure over $\sigma^{(k)}$ which is defined by I. Let $\Pi \subseteq \mathrm{Struct}(\tau)$, $\Sigma \subseteq \mathrm{Struct}(\sigma)$ be problems. Π is $\mathcal{L}$–reducible to Σ if there exist an interpretation I of τ into σ, s.t. for all $\mathfrak{A} \in \mathrm{Struct}(\tau)$, $\mathfrak{A} \in \Pi$ iff $I(\mathfrak{A}) \in \Sigma^{(k)}$ where k is the arity of I. By restricting the logic $\mathcal{L}$ for the interpretations, we obtain low-level reductions: A *quantifier-free reduction* [30] is a reduction whose defining formulas are quantifier-free.

Logic Programs. A logic program is a finite collection of clauses (also called *rules*) of the form

$$A \leftarrow B_1, \ldots, B_m \tag{1}$$

where A is an atom and the B_i's are literals in a first-order language. We use $\neg$ for the negation (as failure) operator in logic programming. A program is positive, if no negation occurs in it.

The meaning of a positive logic program P has been defined in terms of its least model; for logic programming, a number of different semantics have been proposed, since the definition of negation is nontrivial and allows for alternative approaches, cf. [2,8]. For background, consult [35].

Example 1. In the following program, the extension of TC in the least fixpoint is the transitive closure of the binary relation E. We say that the program *computes* TC.

$$TC(x, y) \leftarrow E(x, y)$$
$$TC(x, y) \leftarrow TC(x, z), E(z, y)$$

Generalized Quantifiers (GQs). Let C be a class of logical structures over a relational signature $\sigma = (R_1, \ldots, R_n)$ with arities $a_1, \ldots, a_n$, and suppose that C is closed under isomorphism, i.e., if $\mathfrak{A} \cong \mathfrak{B}$ and $\mathfrak{A} \in C$, then $\mathfrak{B} \in C$. Each such class C has an associated GQ Q_C, whose intended semantics is to check if a relation defined by the underlying logic belongs to C.[3]

---

[3] In logic programming, closedness under isomorphism assures that terms are treated as completely uninterpreted by the GQ. If only particular interpretations are considered, e.g., Herbrand interpretations, weaker closure conditions are in order; cf. the discussion about genericity in logical database query languages.

The extension $\mathcal{L}(Q_C)$ of a logic $\mathcal{L}$ by a GQ Q_C is the closure of $\mathcal{L}$ under the following rule:

If $\varphi_1(\mathbf{x}_1), \ldots, \varphi_n(\mathbf{x}_n)$ are formulas of $L(Q_C)$ where every φ_i has at least a_i free variables $\mathbf{x}_i$, then $Q_C \mathbf{x}_1 \cdots \mathbf{x}_n[\varphi_1, \ldots, \varphi_n]$ is a formula of $L(Q_C)$, in which the occurrences of $\mathbf{x}_1, \ldots, \mathbf{x}_n$ are bound. For clarity, we shall often write the list of remaining free variables $\mathbf{y}$ after the formula.

The semantics of Q_C is defined as follows. Consider the formula

$$\Theta = Q_C \mathbf{x}_1 \cdots \mathbf{x}_n[\varphi_1, \ldots, \varphi_n](\mathbf{y}),$$

and let $\mathfrak{A}$ be a structure and $\mathbf{b}$ a tuple in $\mathfrak{A}$ matching the arity of $\mathbf{y}$. Then $(\mathfrak{A}, \mathbf{b}) \models \Theta$, if and only if the structure $(A, \varphi_1^{\mathfrak{A}, \mathbf{b}}, \ldots, \varphi_n^{\mathfrak{A}, \mathbf{b}})$ belongs to C, where $\varphi_i^{\mathfrak{A}, \mathbf{b}} = \{\mathbf{a} \mid \mathfrak{A} \models \varphi_i[\mathbf{a}, \mathbf{b}]\}$. The *family* $\mathbf{C}$ *of C-quantifiers* is the set $\{Q_C, Q_{C^{(2)}}, Q_{C^{(3)}}, \ldots\}$.

Example 2. The following GQs will be used in further examples.

1. The *complement quantifier:* $Q_\sim = \{(A, U, \{v\}) \mid v \in A - U\}$
2. The *majority quantifier:* $Q_M = \{(A, U, V) \mid U, V \subseteq A, |U| > |V|\}$.
3. The *modularity quantifier:* $Q_k = \{(A, U) \mid |U| \equiv 0 \pmod{k}\}$.
4. The *isomorphism quantifier:* $Q_\cong = \{(A, E, F) \mid (A, E),$
   (A, F) are isomorphic graphs$\}$.
5. The *transitive closure quantifier:* $Q_{TC} = \{(A, E, \{(u, v)\}) \mid$ there is a path
   from u to v in the graph $E \subseteq A \times A\}$.

Complexity Theory. We assume some familiarity with the basic concepts of complexity theory, such as the classes P, NP, L (logspace), oracle computations, and the polynomial hierarchy (PH) whose classes are Σ_k^p, Π_k^p, and Δ_k^p, $k \geq 1$; consult [20,46] for background.

Intertwined with the levels Σ_k^p, Π_k^p of PH are the classes Θ_k^p [63] ($\Sigma_k^p \subseteq \Theta_{k+1}^p \subseteq \Sigma_{k+1}^p$), which are now considered intermediate constituents of PH. The Θ_{k+1}^p classes have natural complete problems in terms of optimization variants of Σ_k^p-complete problems.

The class $\mathrm{P}_{\|}^{\mathrm{NP}}$ is the class of problems decidable by polynomial time Turing machines whose oracle queries are first collected, and then all queries are forwarded to the oracle together. Thus, the queries are independent of each other.

Proposition 1. [63] $\Theta_{k+1}^p = \mathrm{P}_{\|}^{\Sigma_k^p} = \mathrm{L}^{\Sigma_k^p}$.

3   Logic Programs with Generalized Quantifiers

Let τ_0 be a signature for describing the program input, and let τ be an extension of τ_0 by new relational symbols. A *logic program with GQs (GQLP)* on τ is a finite collection $\mathcal{P}$ of rules

$$A \leftarrow B_1, \ldots, B_m \tag{2}$$

where A is a τ-atom whose predicate does not occur in τ_0, and each B_i is either a τ-literal or a generalized quantifier literal (GQ-literal) over τ, which is defined next.

Suppose that Q_C is a GQ defined over the signature $\sigma = (R_1, \ldots, R_n, R_{n+1})$ with associated arities $a_1, \ldots, a_n, a_{n+1} = l$, and that $S_1, \ldots, S_n$ are predicates from τ such that the arity of S_i equals a_i. Then, the formula

$$Q_C \mathbf{x}_1 \cdots \mathbf{x}_{n+1}[S_1(\mathbf{x}_1), \ldots, S_n(\mathbf{x}_n), \mathbf{x}_{n+1} = \mathbf{v}](\mathbf{v}) \tag{3}$$

is a GQ-atom. Notice that the free variables of this GQ-atom are $\mathbf{v} = v_1, \ldots, v_l$. The formula $\mathbf{x}_{n+1} = \mathbf{v}$ defines the singleton relation $\{\mathbf{v}\}$; intuitively, it transfers domain elements into the quantifier. A *GQ-literal* is a possibly negated GQ-atom. For brevity, we denote a GQ atom (3) by

$$Q_C[\mathbf{S}](\mathbf{v}) \qquad (\text{resp. } Q_C[\mathbf{S}], \text{ if } \mathbf{v} \text{ is void}) \tag{4}$$

where $\mathbf{S} = S_1, \ldots, S_n$, and similar for negative GQ-literals.

For any collection $\mathcal{Q}$ of generalized quantifiers, we denote by $\tau_{\mathcal{Q}}^*$ the extension of τ by all predicate letters $Q_C[\mathbf{S}]$, where $Q_C \in \mathcal{Q}$ and $\mathbf{S} = S_1, \ldots, S_n$ is a list of predicate letters S_i from τ which matches the signature of Q_C; every such $Q_C[\mathbf{S}]$ is a *GQ-predicate*. Notice that $\tau_{\mathcal{Q}}^*$ is finite if τ and $\mathcal{Q}$ are finite. Then, a GQLP over τ is syntactically an ordinary logic program over the signature $\tau_{\mathcal{Q}}^*$.

Example 3. Consider the following program $\mathcal{P}$, which uses two GQs: The isomorphism GQ $Q_\cong[G_1, G_2]$, which tells whether G_1 and G_2 are isomorphic graphs, and the transitive closure GQ TC:

$$\begin{aligned} S(x,y) &\leftarrow Q_{TC}[E](x,y), Q_{TC}[E](y,x) \\ G_a(x,y) &\leftarrow E(x,y), S(a,x), S(a,y) \\ G_b(x,y) &\leftarrow E(x,y), S(b,x), S(b,y) \\ Iso &\leftarrow Q_\cong[G_a, G_b] \end{aligned}$$

Suppose that τ_0 contains the relation E and the constant symbols a and b. Given a graph $G = (V, E)$, and vertices $a, b \in V$, this program assigns the propositional letter Iso true on E if the strongly connected components in which a and b lie are isomorphic.

Some remarks about the use of GQs as proposed above are in order. In first-order logic, the scope of a GQ $Q_C \mathbf{x}_1 \cdots \mathbf{x}_n[\varphi_1, \ldots, \varphi_n](\mathbf{v})$ may contain arbitrary first-order formulas φ_i and also GQs. In the context of logic programming, there is no need for allowing complex argument formulas φ_i, since such a φ_i can be associated with a defining equivalent auxiliary relation A_{φ_i} which is computed within the logic program. Moreover, the restriction of free variables to the formula $\mathbf{y} = \mathbf{v}$ is not severe; roughly, free variables $\mathbf{z}$ in a formula $\varphi(\mathbf{x}_i, \mathbf{z})$ can be treated as bound and newly introduced free variables $\mathbf{z}'$ (put in $\mathbf{v}$) can be used to single out the relation $\varphi^{\mathfrak{A}}$ for the semantical definition of the GQ.

In the context of first-order logic, the evaluation of formulas with GQs can be easily defined by structural induction. However, in logic programming, we have to deal with recursion, which makes the definition of a semantics nontrivial, even if no GQs are present, but just negation.

Since negation is provided by the family of complement GQs $Q_\sim$, any semantics of positive logic programs with GQs implicitly provides a semantics for logic programs with negation. Since there is no general consensus about the semantics of a logic program with negation, a generally accepted semantics for GQLPs may not be expected.

We define the semantics of a GQLP $\mathcal{P}$ in the spirit of the stable model semantics for logic programs with negation [21], which is widely accepted. Informally, a stable model of a program $\mathcal{P}$ amounts to a solution of a fixpoint equation

$$\mathfrak{M} = f(\mathcal{P}, \mathfrak{M}),$$

where $f(\mathcal{P}, \mathfrak{M})$ is the result of $\mathcal{P}$ if negation is evaluated with respect to $\mathfrak{M}$.

Our approach is to treat GQ-literals like negative literals; since GQs subsume negation, negation is implicitly treated with (unnegated) GQ-atoms. Under an assumption, based on $\mathfrak{M}$, about the truth of GQ-literals and negative standard literals, a GQLP $\mathcal{P}$ is reduced similar as in [21] to an ordinary positive logic program which is evaluated. If the least minimal model of the latter program justifies the assumptions made by $\mathfrak{M}$, it is considered an acceptable model of $\mathcal{P}$; we call such a model *GQ-stable*.

We next provide a formalization of this concept. Suppose that we have signatures τ_0 and τ as above, a GQLP $\mathcal{P}$ and a structure $\mathfrak{A} \in \text{Struct}(\tau)$.

Then, the *ground instantiation* of $\mathcal{P}$ on $\mathfrak{A}$, denoted $ground(\mathcal{P}, \mathfrak{A})$, is the collection of all interpreted rules $C\vartheta$, where C is from $\mathcal{P}$ and ϑ is any ground substitution over $\mathfrak{A}$.

Definition 1. *Let $\mathcal{P}$ be a GQLP and $\mathfrak{A} \in \text{Struct}(\tau)$. The reduct of $\mathcal{P}$ w.r.t. $\mathfrak{A}$, denoted $red(\mathcal{P}, \mathfrak{A})$, is the set of rules obtained from $ground(\mathcal{P}, \mathfrak{A})$ as follows.*

1. *Remove every rule r with a literal L in the body such that $\mathfrak{A} \not\models L$, where L is either negative or a GQ-literal.*
2. *Remove all negative literals and GQ-literals from the remaining rules.*

Notice that $red(\mathcal{P}, \mathfrak{A})$ is a collection of interpreted Horn clauses; thus, there is a least structure $\mathfrak{B}$, denoted $\mathfrak{A}_\infty(\mathcal{P})$, such that $\mathfrak{B}|\tau_0 = \mathfrak{A}|\tau_0$ (i.e., $\mathfrak{B}$ provides the same input to $\mathcal{P}$ as $\mathfrak{A}$) and $B \models r$, for every rule $r \in red(\mathcal{P}, \mathfrak{A})$. Since $red(\mathcal{P}, \mathfrak{A})$ is an ordinary logic program, the structure $\mathfrak{A}_\infty(\mathcal{P})$, also called the least model of $\mathcal{P}$ w.r.t. $\mathfrak{A}$, can be obtained as the least fixpoint of a monotonic operator, cf. [35].

Definition 2. *Let $\mathcal{P}$ be a GQLP and let $\mathfrak{A}_0 \in \text{Struct}(\tau_0)$. An expansion $\mathfrak{A} \in \text{Struct}(\tau)$ of $\mathfrak{A}_0$ is a GQ-stable model of $\mathcal{P}$ on $\mathfrak{A}_0$, iff it satisfies the fixpoint equation*

$$\mathfrak{A} = \mathfrak{A}_\infty(\mathcal{P}).$$

The collection of all stable models of $\mathcal{P}$ is denoted by $SM(\mathcal{P}, \mathfrak{A}_0)$.

The meaning *of $\mathcal{P}$ on $\mathfrak{A}_0$, denoted $\mathcal{M}_{\mathcal{P}}^{st}(\mathfrak{A}_0)$, is the structure which is the intersection of all GQ-stable models of $\mathcal{P}$ on $\mathfrak{A}_0$, i.e.,*

$$\mathcal{M}_{\mathcal{P}}^{st}(\mathfrak{A}_0) = \bigcap_{\mathfrak{A} \in SM(\mathcal{P}, \mathfrak{A}_0)} \mathfrak{A}.$$

(If $SM(\mathcal{P}, \mathfrak{A}_0) = \emptyset$, then $\mathcal{M}_{\mathcal{P}}^{st}(\mathfrak{A}_0)$ is the unique maximal structure $\mathfrak{B}$ such that $\mathfrak{B}|_{\tau_0} = \mathfrak{A}_0$.)

Example 4. Consider the following program $\mathcal{P}$, which uses the majority quantifier Q_M and the modularity quantifier Q_2 (i.e., the even quantifier).

$$\begin{aligned}
Q(x) &\leftarrow \neg S(x) \\
S(x) &\leftarrow \neg Q(x) \\
A(x) &\leftarrow Q_M[Q, S], S(x) \\
Q(x) &\leftarrow Q_2[A], S(x) \\
W(a, b) &\leftarrow
\end{aligned}$$

Suppose that τ_0 contains merely the constant symbols a and b, and τ contains in addition the relation symbols W, A, Q, and S.

Intuitively, the first two clauses choose complementary extensions for Q and S; the third clause assures that S implies A, if Q holds on more individuals than S; similarly, the fourth clause assures that S implies Q, if A holds true on an even number of elements.

Consider Herbrand models on τ and let $\mathfrak{M}_1 = \{W(a,b), Q(a), Q(b)\}$. (We use the familiar notation for Herbrand models.) This interpretation is a GQ-stable model of $\mathcal{P}$ (w.r.t. the unique Herbrand model $\mathfrak{M}_0 \in Struct(\tau_0)$). Indeed, the reduct $red(\mathcal{P}, \mathfrak{M}_1)$ consists of the clauses

$$\begin{aligned}
W(a, b) &\leftarrow & A(b) &\leftarrow S(b) \\
Q(a) &\leftarrow & Q(a) &\leftarrow S(a) \\
Q(b) &\leftarrow & Q(b) &\leftarrow S(b) \\
A(a) &\leftarrow S(a)
\end{aligned}$$

Clearly, $\mathfrak{M}_1$ is the least model of this program (w.r.t. $\mathfrak{M}_0$). Another GQ-stable model of $\mathcal{P}$ is $\mathfrak{M}_2 = \{W(a,b), B(a,b), S(a), S(b)\}$. The Herbrand model $\mathfrak{M}_3 = \{W(a,b), B(a,b), Q(a), S(b)\}$ is not a GQ-stable model of $\mathcal{P}$: $red(\mathcal{P}, \mathfrak{M}_3)$ contains the clauses $S(b) \leftarrow$ and $Q(b) \leftarrow S(b)$, which means that the least model of $red(\mathcal{P}, \mathfrak{M}_3)$ contains $Q(b)$. In fact, $\mathfrak{M}_1$ and $\mathfrak{M}_2$ are all GQ-stable models of $\mathcal{P}$ w.r.t. $\mathfrak{M}_0$, i.e., $SM(\mathcal{P}, \mathfrak{M}_0) = \{\mathfrak{M}_1, \mathfrak{M}_2\}$. Therefore,

Let us note some interesting properties of GQ-stable models. The following proposition tells that the term "model" is justified.

Proposition 2. *Let $\mathfrak{M}$ be a stable model of a GQLP $\mathcal{P}$. Then, $\mathfrak{M}$ is a model of $\mathcal{P}$, i.e., $\mathfrak{M} \models \forall^*(\mathcal{P})$, where $\forall^*$ is the universal closure of $\mathcal{P}$.*

Another important property is that GQ-stable models conservatively extend the concept of stable models for ordinary logic programs.

Proposition 3. *Let $\mathcal{P}$ be a GQLP. If $\mathcal{P}$ has no GQ-literals, then $\mathfrak{M}$ is a GQ-stable model of $\mathcal{P}$ iff $\mathfrak{M}$ is a stable model of $\mathcal{P}$.*

Therefore, we refer in the sequel to GQ-stable models simply as stable models.

As Example 4 has shown, a program $\mathcal{P}$ may have multiple stable models on a structure. Inherent nondeterminism in the choice of the rules enables the generation of an enumeration of the universe of a finite structure; we illustrate this in the next example.

Example 5. Consider the following program LP_{succ}, which does not contain generalized quantifiers. Therefore, it behaves according to stable semantics for ordinary logic programs [21]. We shall refer to this program in Section 5.

$$
\begin{aligned}
L(x,y) &\leftarrow \neg L^-(x,y) & S(x,y) &\leftarrow \neg S^-(x,y) \\
L^-(x,y) &\leftarrow \neg L(x,y) & S^-(x,y) &\leftarrow \neg S(x,y) \\
L(x,z) &\leftarrow L(x,y), L(x,z) & S^-(x,y) &\leftarrow L^-(x,y) \\
\bot &\leftarrow L^-(x,y), L^-(y,x), x \neq y & S^-(x,z) &\leftarrow L(x,y), L(y,z) \\
\bot &\leftarrow L(x,x)
\end{aligned}
$$

In this program, L and S are binary relations which intuitively describe a linear order $<$ on the domain and the associated successor relation, respectively; the relations L^- and S^- describe the complements of L and S, respectively.

The propositional atom $\bot$ in the heads of the rules is meant to be a logical constant for falsity. Such a constant can be easily defined in stable semantics, by adding a rule

$$ r : \quad P_\bot \leftarrow \bot, \neg P_\bot $$

where $P_\bot$ is a fresh propositional letter which does not occur elsewhere. Then, in every stable model $\mathfrak{M}$, proposition $\bot$ must be false, since otherwise, the rule r cannot be satisfied under stable negation.

It is easy to see that on any finite universe U, the relation L is in every stable model a linear order on U, and S is the corresponding successor relation. *Therefore, we may assume that in any stable model of a program on a finite structure, the universe is enumerated.* Of course, this does not mean that stable model semantics actually defines an order on the domain, because both S and L are empty in the intersection of all stable models.

4   Application: Modular Logic Programming

In this section, we describe how the formalism of logic programs with generalized quantifiers can be fruitfully applied for defining the semantics of logic programs which refer to program modules. This approach can be found in [16,17], and we recall the definitions from there.

A program module is syntactically described by its interface in terms of input and output parameters. Adapted to the logic programming paradigm, a logic programming module has an interface

$$LP[\mathbf{I}; \mathbf{O}],$$

where the $\mathbf{I} = I_1, \ldots, I_m$ are predicates which are imported into the module (i.e., input predicates) and the $\mathbf{O} = O_1, \ldots, O_n$ are predicates which are exported (i.e., output predicates). If we describe the semantics of a module at the extensional level and restrict the number of output relations to $n = 1$, it is clear that such a module is a generalized quantifier. The restriction to a single output relation is not essential, however.

Thus, a module $LP[\mathbf{I}; O_1]$ can be abstractly seen as a GQ $Q_{LP}[\mathbf{I}](\mathbf{x})$, where the length of $\mathbf{x}$ is the arity of O_1. The semantics of modular logic programs is then inherited from the semantics of GQLPs. Notice that this approach is very flexible, and allows for easy incorporation of external predicates and functions, e.g., averaging and counting predicates.

4.1   Syntax of Modular Logic Programs

The syntax of *modular logic programs* (*MLPs*) is the one of GQLPs, with the difference that the GQ-literals are intended to refer to a logic program, which is a *logic program module*. The similarity type of $LP[\mathbf{Q}]$ is the list of arities of predicates in $\mathbf{Q}$.

We shall refer to the calling program as the *main program*, and the called module as the *subprogram*; the GQ-literals in a MLP are termed *call literals*, and the GQ-predicates *call predicates*. An atom is a *call atom*, if its predicate is a call predicate. To distinguish ordinary predicates, atoms, and literals from call predicates, call atoms, and call literals, we call the former *standard* predicates (atoms, literals respectively.)

Definition 3 ([16]). *A logic program module is a pair* $\mu = (LP[\mathbf{Q}], P)$ *of a module head* $LP[\mathbf{Q}]$, *which has an associated integer* $n \geq 0$ *(the arity), and an ordinary logic program* P *(the body), in which the predicates* $\mathbf{Q}$ *are the input predicates and* LP *is the output predicate having arity* n; *syntactically, occurrences of the predicates* $\mathbf{Q}$ *in* P *are restricted to rule bodies. We require that each* LP *module is uniquely identified by its name* LP *and the list of the arities of the* Q_i *in* $\mathbf{Q}$ *(its similarity type).*

In this definition, the predicates $\mathbf{Q}$ can be seen as variables for relations, which are substituted at a call by relations over the domain. All relations in a module $LP[\mathbf{Q}]$ are local; no relations from outside are visible inside a module. The similarity type of a module $LP[\mathbf{Q}]$ is the input signature τ_0 of its implementation program P. Here, τ_0 is relational, and we do for simplicity not consider possible constants and functions in the body P; they could be provided in an extended signature τ_0.

Note that the program P does not contain call literals. Hence, the stable semantics (as well as other semantics) of ordinary logic programming can be readily applied on P. This restriction can be removed and call- resp. GQ-literals may be admitted, provided that some basic constraints are obeyed; we shall address this issue in Section 7.

The relations in P are not visible from the main program, and vice versa. Thus, relations are imported to the subprogram only if they appear in the scope of a call literal, and exported only as the extension of a call literal.

Definition 4 ([16]). *A modular logic program (MLP) is a finite collection LP of rules*

$$A \leftarrow L_1, \dots, L_m$$

where A is a standard atom and the L_i are literals, plus a collection C of logic program modules such that for each call literal $(\neg)LP[\mathbf{Q}](\mathbf{t})$ occurring in P, there is a module $LP[\mathbf{Q}']$ in C where LP has the arity of $\mathbf{t}$ and each $Q_i \in \mathbf{Q}$ has the arity of $Q'_i \in \mathbf{Q}'$.

4.2   Semantics of Modular Logic Programs

As discussed above, we can see LP modules as generalized quantifiers. Thus, by exploiting this correspondence, we can define the semantics of MLPs in terms of the semantics of GQLPs.

Every LP module $\mu = (LP[\mathbf{Q}], P)$ can be seen under the semantics $\mathcal{M}^{st}$ as a GQ $Q_{C(\mu)}$ which is associated with the collection $C(\mu)$ of all structures $\mathfrak{A} = (A, \mathbf{Q}', R')$, where the $\mathbf{Q}'$ are relations for the predicates $\mathbf{Q}$ on A and $R' = \{\mathbf{a}\}$, for any tuple $\mathbf{a}$ over A such that P derives the atom $LP(\mathbf{a})$ on $\mathfrak{A}_0 = (A, \mathbf{Q}')$; i.e.,

$$C(\mu) = \{(A, \mathbf{Q}', \{\mathbf{a}\}) \mid \mathcal{M}^{st}_P((A, \mathbf{Q}')) \models LP(\mathbf{a})\}$$

We call $Q_{C(\mu)}$ the *module quantifier* of μ under $\mathcal{M}^{st}$.

Definition 5. *Let P be an MLP. The meaning of P under $\mathcal{M}^{st}$ is defined as the meaning of P, viewed as GQLP over the collection of GQs associated with the LP modules used by P.*

Example 6. Let $\mathfrak{M} = \{E(1,2), E(2,3), E(3,1), E(2,4), U(1)\}$ be a Herbrand model of the signature $\tau = \{E, U\}$ on the domain $\{1, 2, 3, 4\}$. $E^{\mathfrak{M}}$ is the digraph depicted below.

Consider the module $\mu = (TC[G], P_{TC})$, where P_{TC} is the program from Section 2. The program P_{TC} has on $\mathfrak{A}_0 = \{G(1,2), G(2,3), G(3,1), G(2,4)\}$ the least model $\mathfrak{M}' = \{G(1,2), G(2,3), G(3,1), G(2,4)\} \cup \{TC(i,j), TC(i,4) \mid 1 \leq i,j \leq 3\}$. Thus, a call $TC[E]$ of μ where E is defined in $\mathfrak{M}$, yields that e.g. $TC[E](1,1)$ is true, while $TC[E](4,1)$ is false.

Remark. Certainly, not all generalized quantifiers used have to be program quantifiers. In fact, calls to library modules and external modules, formalized as generalized quantifiers, can be incorporated easily.

5   Expressiveness

Dependency Graph. To obtain a fine-grained analysis of the expressiveness of GQLPs, we generalize the familiar concept of dependency graph from ordinary LPs to GQLPs. For a GQLP LP, we denote by τ_{LP} (or simply τ) the signature of ordinary predicates of LP, and by τ_{LP}^* the extension of τ with the GQ-predicates.

Definition 6 ([16]). *Let LP be a GQLP. The dependency graph of LP, $D(LP)$, is the directed graph constructed as follows. The vertices of $D(LP)$ are the predicates in τ_{LP}^*; an arc from P_1 to P_2 is present in $D(LP)$ if one of 1.–3. applies:*

1. *there is a rule $P_1(\mathbf{t}_1) \leftarrow ..., \neg P_2(\mathbf{t}_2), ...$ in LP.*
2. *P_1 is a call predicate $P[\mathbf{Q}']$ and P_2 occurs in $\mathbf{Q}'$.*
3. *there is a rule $P_1(\mathbf{t}_1) \leftarrow ..., P_2(\mathbf{t}_2), ...$ in LP;*

We color the arc

red,   *if 1. applies and both P_1, P_2 are ordinary predicates;*
yellow, *if P_1 or P_2 is a GQ-predicate;*
green,   *otherwise (i.e., 3. applies and both P_1 and P_2 are ordinary predicates).*

The color of an arc in $D(LP)$ reflects possible nonmonotonic dependency of a P_1 on P_2. A red arc intuitively represents that P_1 is nonmonotonically dependent on P_2, i.e., increasing P_2 decreases P_1; a yellow arc represents that P_1 possibly is nonmonotonically dependent on P_2 (this depends on the properties of the respective module involved); finally, a green arc represents that P_1 is monotonically dependent on P_2.

A predicate P_1 *depends* on P_2, if P_1 reaches P_2 in $D(LP)$. If P_1 and P_2 mutually depend on each other, they are called *mutually recursive*; if neither P_1 depends on P_2 nor vice versa, they are *independent*.

Example 7. Consider a GQLP and its dependency graph. (red arcs are symbolized by label "$\neg$", yellow arcs by "?"; green arcs are unlabeled).

$$E(x,y) \leftarrow G(x,y,y)$$
$$F(x,y) \leftarrow \neg G(x,x,y)$$
$$Y \leftarrow Q_\cong[E,F]$$
$$N \leftarrow \neg Y$$

According to syntactic conditions on the dependency graph $D = D(LP)$, several classes of GQLPs are defined in Table 1.

Table 1. Classes of GQLPs

Class	Symbol	Description
unrestricted	**LP**	arbitrary D
stratified	**Strat**	cycles in D have only green arcs
semistratified	**SStrat**	no cycle in D has a red arc
call recursion free	**CRF**	no cycle in D has a yellow arc
call independent	**CI**	no path in D has two yellow arcs
stratified call independent	**SCI**	cycles in D have only green arcs and no path in D has two yellow arcs
recursion-free	**RF**	D is acyclic
positive	**Pos**	D has no red arcs

Modular Programs. Let $\mathcal{C}_M$ be a class of main programs, and let $\mathcal{C}_S$ be a class of subprograms. Then, $\mathcal{C}_M(\mathcal{C}_S)$ denotes the class of MLPs with main programs from $\mathcal{C}_M$ and subprograms from $\mathcal{C}_S$. (Since subprograms do not contain call predicates, their dependency graphs do not contain yellow edges.)

Expressive Power. Let $\mathcal{L}$ be a class of logic programs, and let C be a complexity class. We say that $\mathcal{L}$ *has expressivity C* (or, equivalently, $\mathcal{L}$ *captures C*) *over a class of structures $\mathcal{K}$*, if

1. for every program LP and structure $\mathfrak{A}$ in $\mathcal{K}$, $\mathcal{M}_{LP}^{st}(\mathfrak{A})$ is computable in C, and
2. every generic (i.e., isomorphism preserving) mapping $f : \mathcal{K} \to \mathcal{K}$ which is computable in C is expressible by a program LP under the semantics $\mathcal{M}^{st}$, i.e., $f = \mathcal{M}_{LP}^{st}$.

Here, computability of a mapping f in C means that deciding whether a relation in $f(\mathfrak{A})$ holds a specific tuple is in C. Semantical restrictions on f may be imposed. E.g., monotone co-NP is the class of all monotone maps f computable in co-NP.

The expressive power of ordinary logic programming over finite structures has been investigated e.g. in [51,15,49]; see [11] for a survey. Stable model semantics was shown to be quite expressive.

Proposition 4 ([51]). **LP** *captures* co-NP, *or equivalently, all Π_1^1 definable queries over finite structures.*

Thus, stable model semantics has the full power of Π_1^1. In fact, it has been shown that even monotone programs can be syntactically captured:

Theorem 1 ([57]). *Monotone NP is captured by the fragment of second order logic where all relational symbols (but not the second order relational variables) occur only positively.*

5.1   Finite Structures

In this subsection, we survey the expressiveness of modular logic programs over finite structures. The results are compactly summarized in Table 2 (**LP^k** and **LP*** are defined in Section 7).

Table 2. Expressiveness of MLP over finite structures

Language	Expressiveness	
LP(X), CI(X), CRF(X) **X** $\in \{\mathbf{LP}, \mathbf{Mon}\}$ **2Pos(LP), SStrat(LP)**	Π_2^1	
2Pos(Mon)	Δ_2^1	*
SCI(X), Strat(X), RF(X) **X** $\in \{\mathbf{LP}, \mathbf{Mon}\}$ **SStrat(Mon)**	Θ_2^1	*
LP, Pos(LP)	Π_1^1	
Pos(Mon)	monotone Π_1^1	
LP^k	Π_k^p	
LP*	Second-Order Logic	

In this table, an asterisk "*" indicates that the proof requires an enumeration of the universe, and thus establishes capturing over ordered finite structures. Note, however, that capturing of Θ_2^p over all finite structures is not possible unless the exponential Boolean hierarchies EBH and ExpBH collapse to some intermediate level; similar results hold for Δ_2^p, cf. [12].

Arbitrary Program Modules. The following theorem states the full expressive power of the formalism.

Theorem 2. LP(LP) *captures Π_2^p over finite structures.*

Proof.   (Sketch) A straightforward algorithm shows that the expressivity of **LP(LP)** is in Π_2^p.

For the other direction, consider a Π_2^1 formula $\varphi(x_1, \ldots, x_n)$ over signature $\tau = (\mathbf{Q})$. In the language extended by a new relation symbol B of arity n, it follows by the extension of Fagin's Theorem [58] that there exists a Π_2^1 formula $\forall R \exists S.\varphi$ which is true on a structure $\mathfrak{A}$ iff $\varphi^{\mathfrak{A}} \subseteq B^{\mathfrak{A}}$. Since stable semantics captures co-NP, there exists a program quantifier P_S for the relation $\neg \exists S.\varphi$. Then the program

$$
\begin{aligned}
B(\mathbf{x}) &\leftarrow \neg B^-(\mathbf{x}) \\
B^-(\mathbf{x}) &\leftarrow \neg B(\mathbf{x}) \\
R(\mathbf{y}) &\leftarrow \neg R^-(\mathbf{y}) \\
R^-(\mathbf{y}) &\leftarrow \neg R(\mathbf{y}) \\
\bot &\leftarrow P_S[\mathbf{Q}, B, R] \\
A(\mathbf{x}) &\leftarrow B(\mathbf{x})
\end{aligned}
$$

computes precisely the relation A as the *intersection* of all stable models.   □

Remark. Note that in **LP(LP)**, the subprograms are arbitrary, so that any quantifiers in co-NPcan be computed, and in particular, arbitrary vectorizations of a given quantifier.

The full expressive power of modular logic programs is available in some of the syntactical fragments for main programs from above.

Corollary 1. **CRF(LP)** *and* **CI(LP)** *capture* Π_2^p *over finite structures.*

Proof.   Follows from the fact that the program in the proof of Theorem 2 is call recursion free and call independent.   □

Corollary 2. **SStrat(LP)** *captures* Π_2^p *over finite structures.*

Proof.   It suffices to show that the complement quantifier $C_{\sim}^{(k)}$ (the vectorization of $C_{\sim}$ to arbitrary arity k) can be expressed in **SStrat(LP)**: To this end, we use additional modules N^k, $k \geq 1$:

$$
\begin{aligned}
N^k[S] \quad &: \\
N^k(\mathbf{x}) &\leftarrow \neg S(\mathbf{x})
\end{aligned}
$$

It appears that the expressive power of the remaining syntactic fragments for main programs have expressive power limited to a subclass of Δ_2^p; on ordered structures, this subclass is Θ_2^p. For the next results, we make use of the following normal form, which has been introduced in [53,54] and generalized in [22].

Theorem 3 (Stewart Normal Form [53,54,22]). *Let* Ω *be NP-complete under quantifier-free reductions over ordered finite structures. Then* Θ_2^p *is captured over ordered structures by the class of formulas*

$$
\exists \mathbf{x}. \ \Omega\mathbf{x}.[\varphi_1](\mathbf{v}) \wedge \neg \Omega\mathbf{x}.[\varphi_2](\mathbf{v})
$$

where φ_1 *and* φ_2 *are quantifier-free formulas.*

Theorem 4. SCI(LP) *captures Θ_2^p over all ordered finite structures.*

Proof. (Sketch) Call-Independence means that the oracle calls can be made in parallel, and hence Theorem 1 implies that the expressivity is in $P_{\|}^{NP}$. On the other hand, Theorem 3 leads to a program of the form

$$A(\mathbf{y}) \leftarrow P_{P1}[\mathbf{Q}](\mathbf{x}, \mathbf{y}), \neg P_{P2}[\mathbf{Q}](\mathbf{x}, \mathbf{y}) \tag{5}$$

$\square$

We say that a language $C(D)$ *obeys the Stewart Normal Form*, if the main programs in $C(D)$ can be written in the form (5). Surprisingly, the expressive power does not increase if we move from **SCI(LP)** to **Strat(LP)**.

Theorem 5. *Over ordered finite structures,* **SCI(LP)** $=$ **Strat(LP)** $=$ **RF(LP)**.

Proof. (Sketch) Every stratified program P can be partitioned into a finite number k of strata $S_1, \ldots, S_k$. Each stratum S_i can be considered as a **SCI(LP)** program over the predicates of the lower strata S_j, $j < i$, and thus can be evaluated by a Θ_2^p machine M_i. Hence, the evaluation of the program can be done by the composition $M_1 \circ \cdots \circ M_k$ of the machines M_i. From the characterization of Θ_2^p in terms of parallel queries (Proposition 1) it follows that this composed computation is in Θ_2^p. This proves that all mappings in **Strat(LP)** are in Θ_2^p. Expressiveness for Θ_2^p over ordered databases follows from Theorem 4.

The result for **RF(LP)** follows immediately from the result about **Strat(LP)** and the fact that the program in the proof of Theorem 4 is actually also from this class. $\square$

Corollary 3. SCI(LP), Strat(LP), *and* **RF(LP)** *obey the Stewart Normal Form.*

Monotone Program Modules. Let us now turn to monotone modules, i.e., programs which define a monotone generalized quantifier. Here, we have the following.

Theorem 6 ([17]). **Pos(Mon)** $=$ **Mon**, *i.e.* **Pos(Mon)** *captures the class of all monotone coNP queries.*

The intuitive reason for this result is that any program in **Pos(Mon)** is monotone, and its evaluation can be defined in terms of a monotone operator [17].

Let us now proceed to richer fragments of main programs than **Pos**. For stratified program, we obtain the following result.

Theorem 7. *Over ordered finite structures,* **Strat(Mon), SStrat(Mon),** *and* **RF(Mon)** *capture* Θ_2^p.

Proof. (Sketch) With monotone modules, all yellow edges become "monotone", and thus all cycles in the given program classes become monotone, and therefore by Theorem 6 can be replaced by single calls to a monotone module. Collapsing the cycles into single calls we obtain a **RF(Mon)** program. Since **RF(LP)** is in Θ_2^p, membership follows.

For expressiveness, we use the Stewart Normal Form again, however we are now confined to monotone modules. By the following lemma, we can overcome this problem by including non-recursive rules which simulate the quantifier-free reductions.

Lemma 1. *Over ordered finite structures, there exists a subclass* **R** *of* (**CI** ∩ **RF**), *such that* **R(Mon)** = **LP**.

To see the lemma, from [56,57] we know that there exist monotone problems which are complete for NP under quantifier-free reductions with successor; such reductions can be easily computed by recursion-free programs.

Thus, by the lemma, it suffices to add a rule like in the proof of Theorem 4, and we obtain an equivalent program in **RF(Mon)**.   □

For the remaining fragments of main programs, the expressiveness is higher.

Corollary 4. **LP(Mon)**, **CRF(Mon)**, *and* **CI(Mon)** *capture* Π_2^p.

Proof. (Sketch) Membership follows from Theorem 2. For expressiveness, we extend the idea of the proof of Theorem 7. Thus, we employ the program used in the proof of Theorem 2, and provide the **LP** modules by Lemma 1.

Since the proof of Lemma 1 requires ordered domains, we add the rules of program LP_{succ} (cf. Section 3) which in every stable model defines a successor relation on the domain. Since the quantifier-free reduction can be simulated recursion-free, the result follows.   □

5.2 Modules under Brave and Skeptical Stable Semantics

The program classes **2Pos(*)** are natural generalizations of the respective **Pos(*)** classes. In the class **2Pos(*)**, it is possible to call a program module in one of two modes, which correspond to the brave and the skeptical variant of the stable model semantics. The mode corresponding to the skeptical variant is just the one that we used for evaluating modules under the stable semantics so far; the mode corresponding to the brave variant follows a dual definition, which takes the union of all stable models rather than their intersection. The mode of a module call $LP[\mathbf{Q}](\mathbf{t})$ is indicated by prefixing the call predicate $LP[\mathbf{Q}]$ with "□" for the skeptical mode, and with "◇" for the brave mode. Similar reasoning modes are employed in epistemic logic programming [4].

This approach supports programming with modules under different calling (or reasoning) modes.

Clearly, **2Pos(LP)** and **LP(LP)** have the same expressive power since negation can be provided by a module (cf. proof of Corollary 2). What happens with

2Pos(Mon) ? The brave modality can be seen as a kind of double negation, similar to $\Diamond = \neg\Box\neg$ in modal logic. Therefore, brave calls to a monotone module are monotone; this means that **2Pos(Mon)** programs amount to positive LPs with monotone GQs; for such programs, a least fixpoint computation can be applied for evaluating the semantics. The expressiveness must be higher than co-NP, since both co-NP-hard and NP-hard modules can be employed.

To investigate the exact expressibility, we utilize the following strong result.

Theorem 8 ([55]). *Let $\Omega \in \mathrm{NP}$ be monotone such that the Hamiltonian Path Problem HP is monotone projection-reducible to Ω. Then Δ_2^p is captured over ordered structures by least fixed point logic, extended by the family of generalized quantifiers for Ω.*

Moreover, every Δ_2^p property can be defined by a formula

$$\mathrm{LFP}_{\mathbf{x},R}\ \Omega\mathbf{y}[\varphi_1](\mathbf{x}, \mathbf{max}) \vee \neg\Omega\mathbf{z}[\varphi_2](\mathbf{x}, \mathbf{max})$$

where R occurs only positively in the formula, and $\mathbf{max}$ denotes a tuple of the maximum elements of the order.

Theorem 9. 2Pos(Mon) *captures Δ_2^p over ordered finite structures.*

Proof.   (Sketch) Since the domain is ordered, we may without loss of generality suppose that constants $0, max$ are available. Membership in Δ_2^p follows from the fact that the modules are monotone by assumption, and therefore, ordinary inflationary evaluation of the main program is possible. Since the closure ordinal is polynomial in the domain size, a polynomial time algorithm with an NP oracle thus obviously exists.

For expressiveness, we apply Theorem 8 to conclude that there exist two modules P_{H1} and P_{H2} which compute H_1 and H_2, under brave and cautious semantics, respectively. Now the main program becomes fairly easy:

$$R(\mathbf{x}, \mathbf{y}) \leftarrow \Box P_{H1}[\mathbf{Q}, R](\mathbf{x}, \mathbf{y})$$
$$R(\mathbf{x}, \mathbf{y}) \leftarrow \Diamond P_{H2}[\mathbf{Q}, R](\mathbf{x}, \mathbf{y})$$
$$A(\mathbf{y}) \leftarrow R(\mathbf{max}, \mathbf{y})$$

$\Box$

Analogous generalizations **2X(*)** can be defined for the other syntactic fragments **X** of main programs; however, since they have negation, the brave calling mode can be easily emulated by the skeptical mode using auxiliary predicates; the expressive power by allowing calls in brave mode thus remains unchanged. Therefore, we do not consider such generalizations here.

5.3   Beyond Finite Structures

Theorem 2 shows that modular logic programming has the same expressive power as disjunctive logic programming over finite structures [15]. Thus, it appears that modular logic programming can be translated into disjunctive logic programming and vice versa. However, such translations are in general only available for finite structures. As shown recently [14], over infinite Herbrand models (resp. the integers) DLP under the stable semantics can express precisely all Π_1^1 definable relations, and thus has the same expressive power as normal logic programming [51] on these models.

It is easy to see that the proof of Theorem 2 works over infinite Herbrand models resp. the integers as well, and thus establishes that MLP can express all Π_2^1 definable relations over these models. Since, as well-known, there are Π_2^1 definable relations over the integers which are not Π_1^1 definable, it follows that MLP is strictly more expressive than DLP on the integers.

In fact, since the stable semantics expresses all Π_1^1 relations over arbitrary structures [51], the proof of Theorem 2 establishes that modular logic programs express Π_2^1 over arbitrary structures. Thus, modular logic programming is a highly expressive (and complex) formalism in general.

6 Complexity

In this section, we survey the complexity of modular logic programs, and address the data and the program complexity [60] of the various classes of modular logic programs which we have introduced above.

Data Complexity. The *data complexity* of an LP class **C** is the complexity of the mappings $\mathcal{M}_{LP}^{st}$ for fixed programs $LP \in \mathbf{C}$. All entries in Table 2 imply completeness results for the respective complexity classes, because capturing of C by a language $\mathcal{L}$ and the existence of a complete problem in C clearly imply that $\mathcal{L}$ contains some C-complete problem. Note, however, that the converse does not necessarily hold; a language $\mathcal{L}$ can be C-complete while it does not capture C. For example, **Pos** is known to be P-complete with respect to data complexity, while it does not capture P (see [11]). Therefore, capturing of C is a stronger concept than C-completeness.

Program Complexity. Besides data complexity, Vardi [60] introduced the concept of *program (resp. combined) complexity*: program complexity[4] is the complexity of evaluating varying logic programs over a fixed input structure $\mathfrak{A}_0$; in combined complexity, both LP and $\mathfrak{A}_0$ are varying. Typically, program complexity and combined complexity are the same (wrt logspace reducibility), and they are exponentially higher than the data complexity.

In [23,24], a method for deriving program complexity results from expressivity results has been developed. It is based on complexity theoretic investigations of succinct problems, i.e. computational problems, whose instances are not given in the usual way as a string, but by a Boolean circuit or by a formula representing this string. The representation by a formula is of particular interest, since it can be used for simulating succinct input representation within most logics used in computer science.

---

[4] In [60], the term *expression complexity* was used instead of *program complexity*. Both terminologies are widely used.

Table 3. Complexity of MLP over finite structures

Language	Data Complexity	Program Complexity
LP(X), CI(X), CRF(X) $X \in \{\mathbf{LP}, \mathbf{Mon}\}$ **2Pos(LP), SStrat(LP)**	Π_2^p	$E\Pi_2^p = \mathrm{NEXP}^{\mathrm{NP}}$
2Pos(Mon)	Δ_2^p	$E\Delta_2^P = \mathrm{EXP}^{\mathrm{NP}}$
SCI(X), Strat(X), RF(X) $X \in \{\mathbf{LP}, \mathbf{Mon}\}$ **SStrat(Mon)**	Θ_2^p	$E\Theta_2^P = \mathrm{PSPACE}^{\mathrm{NP}}$
LP, Pos(LP)	coNP	coNEXP
Pos(Mon)	coNP	coNEXP
LP^k	Π_k^1	$E\Pi_k^P$
LP*	PH	EXPSPACE

The following two syntactic closure properties are easily satisfied by most logics:

Definition 7. *A logic $\mathcal{L}$ is* uniformly interpretation closed, *if there exists a mapping*

$$(\rho, \varphi_1, \ldots, \varphi_r) \to \rho[R_1/\varphi_1, \ldots, R_r/\varphi_r]$$

which is computable in logspace and such that for all structures $\mathfrak{A}$, $\mathcal{L}$-expressions ρ, and quantifier-free first order formulas $\varphi_1, \ldots, \varphi_r$ it holds that

$$(\mathfrak{A}, \varphi_1^{\mathfrak{A}}, \ldots, \varphi_r^{\mathfrak{A}}) \models \rho \iff \mathfrak{A} \models \rho[R_1/\varphi_1, \ldots, R_r/\varphi_r]$$

Definition 8. *A logic $\mathcal{L}$ is* uniformly vector closed, *if there exists a mapping*

$$(\varphi, w) \mapsto \varphi^{(|w|)}$$

which is computable in logspace and such that

$$\mathrm{Mod}(\varphi)^{(|w|)} = \mathrm{Mod}(\varphi^{(|w|)})$$

for all binary strings w and $\mathcal{L}$-expressions φ.

For brevity we say that a logic is *uniformly closed* if it is both uniformly interpretation closed and uniformly vector closed.

Lemma 2 ([24]). *If $\mathcal{L}$ captures C and is uniformly closed, then $\mathcal{L}$ has program complexity at least $E(C)$ and $Exp(C)$ under logspace-reductions.*

Here, $E(C)$ and $Exp(C)$ refer to the exponential analogues of the class C; thus $E(\mathrm{NP}) = \mathrm{NE}$, and $Exp(\mathrm{NP}) = \mathrm{NEXPTIME}$; see [24] for details.

We apply this lemma for precisely capturing the program complexity of all syntactic fragments of modular logic programs that we have considered. The results are summarized in Table 3.

Theorem 10. *All program complexity results of Table 3 hold.*

Proof. All languages in this paper are uniformly interpretation closed: For a quantifier-free formula φ_i, it is easy to write a program P_φ simulating φ_i, whose output predicate is R_i. Uniform vector closure is obtained by replacing all occurrences of variables x by tuples $x_1, \dots, x_n$ of variables, for $n \geq 1$. □

In fact, the languages in this paper are even able to simulate succinct circuit problems, cf. [24]. Therefore, program complexity can be related to the concept of leaf languages [61,62]; it was shown that leaf languages always admit complete problems under quantifier-free reductions. A straightforward encoding of programs as finite structures shows that the above notions of uniformity can be restricted to quantifier-free uniformity (see [61]), and therefore we obtain the following corollary:

Corollary 5. *Let $\mathcal{L}$ be any LP language considered in this paper. If $\mathcal{L}$ captures C, then $\mathcal{L}$ has program complexity at least $\mathrm{Leaf}^P(C)$. This holds even with respect to quantifier-free reductions.*

Remark. (1) In [41], a diagonalization argument is given to prove that there are arity hierarchies within certain logics. A similar argument shows that the combined complexity of languages closed under complementation is always higher than the expressiveness.

Indeed, consider programs over strings (i.e., unary alphabets with linear order). For a program P, let $\lceil P \rceil$ denote its encoding as a string. For a language $\mathcal{L}$ which is closed under complementation, we define the set $AUTSAT(\mathcal{L}) = \{\lceil P \rceil \mid \lceil P \rceil \models P\}$; i.e., $AUTSAT(\mathcal{L})$ contains those programs P which evaluate to true on their own representation. The complexity of $AUTSAT(\mathcal{L})$ is clearly a lower bound for the combined complexity of $\mathcal{L}$. Moreover, $AUTSAT(\mathcal{L})$ is not definable in $\mathcal{L}$: Suppose its complement $\Delta(\mathcal{L})$ is definable in $\mathcal{L}$: Then there exists a program $P' \in \mathcal{L}$, such that for all programs $P \in \mathcal{L}$, $\lceil P \rceil \models P'$ iff $\lceil P \rceil \notin AUTSAT(\mathcal{L})$. Setting $P = P'$, we arrive at a contradiction.

As a matter of fact, when we say that the data complexity of a language (e.g. NP) is strictly weaker than its program complexity (e.g. NEXPTIME), this is just an application of the time hierarchy theorems [32], which are based on a more refined diagonalization technique.

(2) The same technique can be used to show in an easy way that restricting the arity of relation symbols in those classes of programs which are closed under complement yield infinite hierarchies wrt to expressiveness: Consider for example **RF(LP)**. For $k \geq 1$, $\mathbf{RF}_k(\mathbf{LP}_k)$ is the class of **RF(LP)** programs, where all relations have maximum arity k. It is easy to see that the combined complexity of

$\mathbf{RF}(\mathbf{LP})$ remains in Θ_2^p, and hence there is a $\mathbf{RF}_k(\mathbf{LP}_k)$ program LP computing $\Delta(\mathbf{RF}_k(\mathbf{LP}_k))$. Like in (1), we conclude that LP must have arity at least $k+1$, and thus, the result follows.

Theorem 11. *Let $\mathcal{L}$ be any logic programming language considered in this paper. If $\mathcal{L}$ captures a complexity class C, which is closed under complement, then $\mathcal{L}$ has infinite arity hierarchies wrt to expressiveness.*

7   Nested Program Modules

It is evident that a modular logic program can again be used as a subprogram to a new logic program; proceeding this way, we obtain nested modular programs of arbitrary depth. A program which is built up this way and which does not contain cyclic dependencies of subprograms (i.e., no recursive calls of modules), is called *hierarchical*.

For finitely nested, hierarchical programs, a semantics can be inductively defined, based on the semantics of programs of nesting depth 2 which we presented. In particular, stable models of such programs can be naturally defined, and stratified, positive, etc. fragments of programs can be considered. We do not carry out this here; the formalization should be clear.

We extend out notation for nested classes in the following way. For any $k \geq 1$, $\mathbf{LP}^k$ is the class of all modular logic programs with nesting depth bounded by k, and $\mathbf{LP}^* = \bigcup_{k \geq 1} \mathbf{LP}^k$. In particular, $\mathbf{LP} = \mathbf{LP}^1$ and $\mathbf{LP}^{k+1} = \mathbf{LP}(\mathbf{LP}^k)$.

Theorem 12. $\mathbf{LP^k}$ *captures the class Π_k^p of PH, and $\mathbf{LP}^*$ captures PH.*

Proof.   Analogous to Theorem 2. $\square$

Thus, modular logic programming with nestings is as expressive as full second-order logic over finite structures (cf. [58]). In fact, this is an interesting difference to extensions of first order logic by generalized quantifiers: Although logic programs can be considered as a nested quantifier, there is no collapse to the Θ_2^p level, as there is in extensions of first order logic by NP quantifiers [22]; the reason is that in stable model semantics, there is a direct access to relational variables which is not the case for generalized quantifiers.

Moreover, from the considerations in Section 5.3, we can conclude that nested modular logic programs allow to express full second order logic over arbitrary structures. Thus, $\mathbf{LP}^*$ programming with nestings is a tremendously expressive language.

An inductive definition of different fragments by nesting programs from different fragments of $\mathbf{LP}$ is straightforward. For example, we can define classes such as $\mathbf{Strat}(\mathbf{LP}(\mathbf{Strat}))$, $\mathbf{Strat}(\mathbf{CRF}(\mathbf{LP}))$ etc.

The capturing results from Table 2 can be lifted to all constituent classes of PH by appropriate classes of nested MLPs. For example, $\mathbf{Strat}(\mathbf{SStrat}(\mathbf{LP}))$ captures Θ_3^p, and it obeys the Stewart Normal Form in the main programs. Note that $\mathbf{LP}(\mathbf{Strat}(\mathbf{LP}))$ collapses to $\mathbf{LP}(\mathbf{LP})$ since the logspace computations in the medium module can be transferred to the main program.

8   Conclusion

The extension of logic programs by generalized quantifiers is a useful tool for a natural formalization of modular logic programs.

As shown above, the formalism for modular logic programming described in [17,16] has high expressive capability and equals the power of second order logic. Over finite structures, it captures the polynomial hierarchy, and natural fragments of the language capture the constituent classes in this hierarchy.

Several issues remain for further research. On the semantical side, an intriguing issue is a semantics for nested modular logic programs which are not hierarchical, i.e., modular logic programs which use recursive calls of modules. Furthermore, it would be interesting to experience semantics for modular logic programs which are based on semantics different from the stable model semantics; in particular, for semantics which use multi-valued logics (see [11] for several alternative semantics of logic programs).

Another direction of research is the study of the effect of restricting arities of generalized quantifiers on expressiveness. In particular, it would be interesting to see arity hierarchies for program classes capturing complexity classes which are not closed under complementation (e.g., **LP(LP)**). Moreover, the picture of the arity hierarchies in the other classes of programs remains to be refined.

Acknowledgment

We are grateful to Janos Makowsky for valuable comments and remarks.

References

1. A. Aho and J. Ullman. Universality of Data Retrieval Languages. In *Proceedings ACM Symposium on Principles of Programming Languages*, pages 110–117, 1979.
2. K. Apt and N. Bol. Logic Programming and Negation: A Survey. *Journal of Logic Programming*, 19/20:9–71, 1994.
3. A. Badia, M. Gyssens, and D. V. Gucht. Query languages with generalized quantifiers. In R.Ramakrishnan, editor, *Applications of Logic Databases*, pages 235–258. Kluwer Academic Publishers, 1995.
4. C. Baral and M. Gelfond. Logic Programming and Knowledge Representation. *Journal of Logic Programming*, 19/20:73–148, 1994.
5. J. Barwise. Some Applications of Henkin Quantifiers. *Israel Journal of Mathematics*, 25:47–63, 1976.
6. J. Barwise and R. Cooper. Generalized Quantifiers and Natural Language. *Linguistic and Philosophy*, 4:159–219, 1981.
7. J. Barwise and S. Feferman. *Model-Theoretic Logics*. Springer, 1985.
8. N. Bidoit. Negation in rule-based database systems: a survey. *Theoretical Computer Science*, 78:3–83, 1991.
9. A. Blass and Y. Gurevich. Henkin Quantifiers and Complete Problems. *Annals of Pure and Applied Logic*, 32:1–16, 1986.
10. M. Bugliesi, E. Lamma, and P. Mello. Modularity in Logic Programming. *Journal of Logic Programming*, 19/20:443–502, 1994.

11. E. Dantsin, T. Eiter, G. Gottlob, and A. Voronkov. Complexity and Expressive Power of Logic Programming. In *Proceedings of the Twelth IEEE International Conference on Computational Complexity (CCC '97)*, pages 82–101, 1997.
12. A. Dawar, G. Gottlob, and L. Hella. Capturing Relativized Complexity Classes without Order. *Mathematical Logic Quarterly*, 15:109–122, 1998.
13. H.-D. Ebbinghaus and J. Flum. *Finite Model Theory*. Perspectives in Mathematical Logic. Springer, 1995.
14. T. Eiter and G. Gottlob. Expressiveness of Stable Model Semantics for Disjunctive Logic Programs with Functions. *Journal of Logic Programming*, 33(2):167–178, 1997.
15. T. Eiter, G. Gottlob, and H. Mannila. Disjunctive Datalog. *ACM Transactions on Database Systems*, 22(3):364–417, September 1997.
16. T. Eiter, G. Gottlob, and H. Veith. Modular Logic Programming and Generalized Quantifiers. In J. Dix, U. Furbach, and A. Nerode, editors, *Proceedings of the 4th International Conference on Logic Programming and Nonmonotonic Reasoning (LPNMR-97)*, number 1265 in LNCS, pages 290–309. Springer, 1997.
17. T. Eiter, G. Gottlob, and H. Veith. Modular Logic Programming and Generalized Quantifiers. Technical Report CD-TR 97/111, Christian Doppler Laboratory for Expert Systems, TU Vienna, January 1997. Preliminary Report.
18. H. Enderton. Finite Partially-Ordered Quantifiers. *Zeitschrift für Mathematische Logik und Grundlagen der Mathematik*, 16:393–397, 1970.
19. M. Fitting. Enumeration Operators and Modular Logic Programming. *Journal of Logic Programming*, 4:11–21, 1987.
20. M. Garey and D. S. Johnson. *Computers and Intractability – A Guide to the Theory of NP-Completeness*. W. H. Freeman, New York, 1979.
21. M. Gelfond and V. Lifschitz. The Stable Model Semantics for Logic Programming. In *Logic Programming: Proceedings Fifth Intl Conference and Symposium*, pages 1070–1080, Cambridge, Mass., 1988. MIT Press.
22. G. Gottlob. Relativized Logspace and Generalized Quantifiers over Ordered Finite Structures. *Journal of Symbolic Logic*, 62(2):545–574, 1997.
23. G. Gottlob, N. Leone, and H. Veith. Second-Order Logic and the Weak Exponential Hierarchies. In J. Wiedermann and P. Hajek, editors, *Proceedings of the 20th Conference on Mathematical Foundations of Computer Science (MFCS '95)*, number 969 in Lecture Notes in Computer Science, pages 66–81, Prague, 1995.
24. G. Gottlob, N. Leone, and H. Veith. Succinctness as a Source of Complexity in Logical Formalisms. *Annals of Pure and Applied Logic*, 97:231–260, 1999.
25. E. Grädel. Capturing Complexity Classes with Fragments of Second Order Logic. *Theoretical Computer Science*, 101:35–57, 1992.
26. P. Hájek. *Metamathematics of Fuzzy Logic*. Kluwer Academic Publishers, 1998.
27. L. Henkin. Some Remarks on Infinitely Long Formulas. In *Infinitistic Methods, Proc. Symp. on Foundations of Mathematics*, pages 167–183. Warsaw, Panstwowe Wydawnictwo Naukowe and Pergamon Press, 1961.
28. J. Hintikka. Quantifiers vs. Quantification Theory. *Linguistic Inquiry*, 5:153–177, 1974.
29. J. Hintikka. Quantifiers in Logic and Quantifiers in Natural Languages. In S. Körner, editor, *Philosophy of Logic*, chapter V. Univ. of California Press, Berkeley et al, 1976.
30. N. Immerman. Languages that Capture Complexity Classes. *SIAM Journal of Computing*, 16:760–778, 1987.
31. N. Immerman. *Descriptive Complexity*. Springer, 1997.

32. D. S. Johnson. A Catalog of Complexity Classes. In J. van Leeuwen, editor, *Handbook of Theoretical Computer Science*, volume A, chapter 2. Elsevier Science Publishers B.V. (North-Holland), 1990.

33. M. Krynicky and J. Väänänen. Henkin and Function Quantifiers. *Annals of Pure and Applied Logic*, 43:273–292, 1989.

34. P. Lindström. First Order Predicate Logic with Generalized Quantifiers. *Theoria*, 32:186–195, 1966.

35. J. Lloyd. *Foundations of Logic Programming*. Springer, Berlin, 1984, 1987.

36. J. Makowsky. Model Theory and Computer Science: An Appetizer. In S. Abramsky, D. Gabbay, and T. Maibaum, editors, *Handbook of Logic in Computer Science*, volume I, chapter 6. Oxford University Press, 1992.

37. J. Makowsky. Capturing Relativized Complexity Classes with Lindström Quantifiers. In W. D. Pauli-Schimanovich, editor, *The Foundational Debate*, pages 133–140. Kluwer Academic Publishers, 1995.

38. J. Makowsky and Y. Pnueli. Oracles and Quantifiers. In *Computer Science Logic 1993 (CSL'93)*, number 832 in LNCS, pages 189–222. Springer, 1994.

39. J. Makowsky and Y. Pnueli. Computable Quantifiers and Logics over Finite Structures. In M. Krynicki, M. Mostowski, and L. Szczerba, editors, *Quantifiers: Logics, Models and Computation, Volume I*, pages 313–357. Kluwer Academic Publishers, 1995.

40. J. Makowsky and Y. Pnueli. Logics Capturing Relativized Complexity Classes Uniformly. In D. Leivant, editor, *Proceedings LCC'94, Logic and Computational Complexity*, number 960 in LNCS, pages 463–479. Springer, 1995.

41. J. Makowsky and Y. Pnueli. Arity and Alternation in Second-Order Logic. *Annals of Pure and Applied Logic*, 78:189–202, 1996.

42. D. Miller. A Theory of Modules in Logic Programming. In *Proceedings of the International Logic Programming Symposium (ILPS '86)*, pages 106–114. MIT Press, 1986.

43. R. Montague. In R. Thomasson, editor, *Formal Philosophy: Selected Papers*. Yale University Press, New Haven, Conn, 1974.

44. A. Mostowski. On a Generalization of Quantifiers. *Fundamenta Mathematicae*, 44:12–36, 1957.

45. R. O'Keefe. Towards an Algebra for Constructing Logic Programs. In *Proceedings of the IEEE Symposium on Logic Programming*, pages 152–160. IEEE CS Press, 1985.

46. C. H. Papadimitriou. *Computational Complexity*. Addison-Wesley, 1994.

47. S. Rao, A. Badia, and D. Van Gucht. Processing Queries Containing Generalized Quantifiers. Technical Report TR 428, Computer Science Department, Indiana University, 1996.

48. B. Russell. *The Principles of Mathematics*. George Allen and Unwin, London, 1903.

49. D. Saccá. Deterministic and Nondeterministic Stable Model Semantics for Unbound DATALOG Queries. In *Proceedings 5th Intl. Conference on Database Theory (ICDT-95)*, LNCS 893, pages 353–367, January 1995.

50. G. Sandu and J. Väänänen. Partially Ordered Connectives. *Zeitschrift für Mathematische Logik und Grundlagen der Mathematik*, 38:361–372, 1992.

51. J. Schlipf. The Expressive Powers of Logic Programming Semantics. *Journal of Computer and System Sciences*, 51(1):64–86, 1995. Abstract in Proc. PODS 90, pp. 196–204.

52. J. Spivey. *The Z Notation: A Reference Manual*. Addison Wesley, Harlow, 1992.

53. I. Stewart. Logical Characterizations of Bounded Query Classes I: Logspace Oracle Machines. *Fundamenta Informaticae*, 18:65–92, 1993.
54. I. Stewart. Logical Characterizations of Bounded Query Classes II: Polynomial-Time Oracle Machines. *Fundamenta Informaticae*, 18:93–105, 1993.
55. I. Stewart. Incorporating generalized quantifiers and the least fixed point operator. In *Proc. Computer Science Logic*, number 832 in LNCS, pages 318–333, 1994.
56. I. Stewart. Complete problems for monotone NP. *Theoretical Computer Science*, 145:147–157, 1995.
57. I. Stewart. Logical Descriptions of monotone NP problems. *Journal of Logic and Computation*, 4:337–357, 1995.
58. L. J. Stockmeyer. The Polynomial-Time Hierarchy. *Theoretical Computer Science*, 3:1–22, 1977.
59. J. Väänänen. Generalized Quantifiers, an Introduction. This volume.
60. M. Vardi. Complexity of Relational Query Languages. In *Proceedings 14th STOC*, pages 137–146, San Francisco, 1982.
61. H. Veith. Succinct Representation, Leaf Languages, and Projection Reductions. *Information and Computation*, 142(2):207–236, 1998.
62. H. Vollmer. A General Quantifier Concept in Computational Complexity Theory. This volume.
63. K. Wagner. Bounded Query Classes. *SIAM Journal of Computing*, 19(5):833–846, 1990.
64. W. Walkoe. Finite Partially-Ordered Quantification. *Journal of Symbolic Logic*, 35:535–555, 1970.
65. J. Wordsworth. *Software Engineering with B*. Addison Wesley, 1996.

A Generalized Quantifier Concept in Computational Complexity Theory

Heribert Vollmer

Theoretische Informatik, Universität Würzburg, Am Exerzierplatz 3
D-97072 Würzburg, Germany
vollmer@informatik.uni-wuerzburg.de

Abstract. A notion of generalized quantifier in computational complexity theory is explored and used to give a unified treatment of leaf language definability, oracle separations, type 2 operators, and circuits with monoidal gates. Relations to Lindström quantifiers are pointed out.

1 Introduction

In this paper we develop a unified view at some notions that appeared in computational complexity theory in the past few years. This will be in the form of operators transforming complexity classes into complexity classes. Each such operator is given in the form of a quantifier on strings. This will immediately subsume as special cases the well known universal, existential, and counting quantifiers examined in various complexity theoretic settings [SM73,Wra77,Wag86b,Wag86a] [Tor91]. But also a lot of constructions from other subareas of complexity theory can best be understood in terms of such operators. These include circuits with arbitrary monoidal gates [BIS90,BI94], oracle operators [BW96,BVW96], leaf languages (introduced in [BCS92,Ver93] and examined for different computation models in [HLS+93,JMT94,CMTV96]). We survey some results from these areas and establish some new connections.

In finite model theory, examinations of the expressive power of various logics enhanced by Lindström quantifiers form a very well established field of active research. Descriptive complexity theory has characterized a great bulk of complexity classes by such logics. We will show that classes defined by our general operator can in a uniform way be characterized by model theoretic means using Lindström quantifiers.

In the following, we assume some familiarity of the reader with basic formal language theory (refer to [RS97]), basic complexity classes and resource-bounded reducibilities (refer to the standard literature, e.g. [Pap94,BC94,BDG95]; all complexity classes that appear in this paper without definition are defined in [Joh90]), as well as with the basics of finite model theory (refer to [Vää94,EF95]).

2 Definition

Given a language A over some alphabet Σ, we denote the characteristic function of A by χ_A, i.e. for all $x \in \Sigma^*$, $\chi_A(x) = 1$ if $x \in A$, and $\chi_A(x) = 0$ otherwise.

J. Väänänen (Ed.): ESSLLI'97, LNCS 1754, pp. 99–123, 1999.

We will always assume some order on the alphabets we use; therefore it makes sense to talk about the lexicographic order $\prec$ of Σ^*, and for $x, y \in \Sigma^*$, $x \preceq y$, we define the characteristic string of A from x to y as $\chi_A[x \ldots y] =_{\text{def}}$ $\chi_A(x)\chi_A(x+1)\cdots\chi_A(y)$. Here, $x+1$ denotes the successor of x. In fact, we will presuppose an underlying bijection between Σ^* and the set $\mathbf{N}$ of natural numbers, and we use the notation $\chi_A[i \ldots j]$ for $i, j \in \mathbf{N}$.

Let $\langle \cdot, \cdot \rangle$ denote a standard pairing function. For a set $A \subseteq \Sigma^*$ and a string $x \in \Sigma^*$, define $A_x =_{\text{def}} \{ y \mid \langle x, y \rangle \in A \}$.

Looking now at the well-known characterization of the polynomial hierarchy by polynomially length-bounded universal and existential quantifiers $\exists^p$, $\forall^p$ [Wra77], the following is clear:

- A language L is in NP if and only if there is a language $A \in \mathrm{P}$ and a function f computable in polynomial time such that for all x,

$$x \in L \iff \chi_{A_x}[0 \ldots f(x)] \in (0+1)^*1(0+1)^* \text{ (i.e., contains a ``1'').}$$

- A language L is in coNP if and only if there is a language $A \in \mathrm{P}$ and a polynomial p such that for all x,

$$x \in L \iff \chi_{A_x}[0 \ldots f(x)] \in 1^* \quad \text{(i.e., consists out of ``1''s only).}$$

An analogous result holds for the class PP, which was characterized in [Wag86a] in terms of the so called polynomially length-bounded counting quantifier C^p:

- A language L is in PP if and only if there is a language $A \in \mathrm{P}$ and a function f computable in polynomial time such that for all x,

$$x \in L \iff \chi_{A_x}[0 \ldots f(x)] \text{ contains more ``1''s than ``0''s.}$$

The class US (for unique solution) is defined by polynomial time nondeterministic Turing machines M which accept an input x if and only if there is exactly one accepting path in the computation tree of M on x.

- A language L is in US if and only if there is a language $A \in \mathrm{P}$ and a function f computable in polynomial time such that for all x,

$$x \in L \iff \chi_{A_x}[0 \ldots f(x)] \in 0^*10^*.$$

Thus we see that here the semantics of quantifiers is defined by giving languages over the binary alphabet ($E =_{\text{def}} (0+1)^*1(0+1)^*$ for $\exists^p$, $U =_{\text{def}} 1^*$ for $\forall^p$, and $\mathrm{maj} =_{\text{def}} \{ w \in \{0,1\}^* \mid w \text{ contains more ``1''s than ``0''s} \}$ for C^p). The following generalization now is immediate:

Let $B \in \{0,1\}^*$, let $\mathcal{K}$ be a class of sets, and $\mathcal{F}$ be a class of functions from $\Sigma^* \to \mathbf{N}$. Define the class $(B)^{\mathcal{F}}\mathcal{K}$ to consists of all sets L for which there exist some $A \in \mathcal{K}$ and some function $f \in \mathcal{F}$ such that for all $x \in \Sigma^*$, $x \in L \iff \chi_{A_x}[0 \ldots f(x)] \in B$.

We use the following shorthands: Write $(A)^\mathrm{P}\mathcal{K}$ $((A)^{\log}\mathcal{K}, (A)^{\mathrm{plog}}\mathcal{K}$, resp.), if $\mathcal{F}$ is the class of all functions from $\Sigma^* \to \mathbf{N}$ computable in polynomial time (logarithmic time, polylogarithmic time, resp.) on deterministic Turing machines (i.e., $\mathcal{F} = \mathrm{FP}$, $\mathcal{F} = \mathrm{FDLOGTIME}$, $\mathcal{F} = \mathrm{FPOLYLOGTIME}$, resp.). For sub-linear time bounds we use Turing machines with index tape and random access to their input, working in the unrestricted mode (for background, refer to [CC95,RV97]). Observe that a function $f \in \mathrm{FP}$ is polynomially length-bounded, $f \in \mathrm{FDLOGTIME}$ is length-bounded by some function $c \cdot \log n$, and $f \in \mathrm{FPOLYLOGTIME}$ is polylogarithmically length-bounded. If $\mathcal{L}$ is a class of languages, then $(\mathcal{L})^{\mathcal{F}}\mathcal{K} =_{\mathrm{def}} \bigcup_{B \in \mathcal{L}} (B)^{\mathcal{F}}\mathcal{K}$.

If we take the above three languages E, U, and maj, and look at different function classes $\mathcal{F}$, we get the existential, universal, and counting quantifier for various length-bounds.

The above definition appeared in [Vol96b] and (for the special case $\mathcal{F} = \mathrm{FP}$) in [BS97].

3   Polynomial Time Leaf Languages

The most examined special case of our general operator is probably the polynomial time case, i.e. the base class $\mathcal{K}$ is the class P (and $\mathcal{F} = \mathrm{FP}$). In this case there is a very intuitive way of visualizing the operator via so called *leaf languages*.

3.1   Definition

In the leaf language approach to the characterization of complexity classes, the acceptance of a word input to a nondeterministic machine depends only on the values printed at the leaves of the computation tree. To be more precise, let M be a nondeterministic Turing machine, halting on every path printing a symbol from an alphabet Σ, with some order on the nondeterministic choices. Then, leafstring$^M(x)$ is the concatenation of the symbols printed at the leaves of the computation tree of M on input x (according to the order of M's paths given by the order of M's choices). Given now a language $B \subseteq \{0,1\}^*$, we define $\mathrm{Leaf}^M(B) = \{\, x \mid \mathrm{leafstring}^M(x) \in B \,\}$.

Call a computation tree of a machine M *balanced*, if all of its computation paths have the same length, and moreover, if we identify every path with the string over $\{0,1\}$ describing the sequence of nondeterministic choices on this path, then there is some string z such that all paths y with $|y| = |z|$ and $y \preceq z$ (in lexicographic ordering) exist, but no path y with $y \succ z$ exists.

A leaf language $B \subseteq \Sigma^*$ now defines the class $\mathrm{BLeaf}^\mathrm{P}(B)$ of all languages L for which there exists a nondeterministic polynomial time machine M whose computation tree is always balanced, such that $L = \mathrm{Leaf}^M(B)$. Let $\mathcal{C}$ be a class of languages. The class $\mathrm{BLeaf}^\mathrm{P}(\mathcal{C})$ consists of the union over all $B \in \mathcal{C}$ of the classes $\mathrm{BLeaf}^\mathrm{P}(B)$.

This computation model was introduced by Bovet, Crescenzi, and Silvestri, and independently Vereshchagin [BCS92,Ver93] and later examined by Hertrampf, Lautemann, Schwentick, Vollmer, and Wagner [HLS+93], and Jenner, McKenzie, and Thérien [JMT94], among others. See also the textbook [Pap94, pp. 504f].

Jenner, McKenzie, and Thérien also considered the case where the computation trees are not required to be balanced. For that case, let B be any language. Then, the class $\text{Leaf}^P(B)$ consists of those languages L for which there exists a nondeterministic polynomial time machine M without further restriction, such that $L = \text{Leaf}^M(B)$. Let $\mathcal{C}$ be a class of languages. The class $\text{Leaf}^P(\mathcal{C})$ consists of the union over all $B \in \mathcal{C}$ of the classes $\text{Leaf}^P(B)$. (Strictly speaking, the definition of *balanced* given in [JMT94] is different from ours and, at first sight, slightly more general. However, it is easy to see that both definitions are equivalent.)

The reader now might wonder about the seemingly unnatural condition that the nondeterministic choices of M are ordered. In fact, most complexity classes of current focus can be defined without this assumption—in this case the leaf language B has the special property that we can permute the letters in a given word without affecting membership in B. (Cf. our results on *cardinal languages* in Sect. 3.3 below.) However, strange classes where the order of the paths is important for their definition are conceivable, and the results presented below, especially the oracle separation criterion (Theorem 6), also hold for these pathologic cases.

The following connection to our generalized quantifier now is not too hard to see.

Theorem 1. *Let $B \subseteq \{0,1\}^*$. Then $(B)^P P = \text{BLeaf}^P(B)$.*

Proof sketch. ($\subseteq$) Let $L \in (B)^P P$, $x \in L \iff \chi_{A_x}[0 \ldots f(x)] \in B$. The nondeterministic machine, given x, branches on all possible second inputs y in the range $0, \ldots, f(x)$, and outputs $\chi_A(x,y)$.

($\supseteq$) Let $L \in \text{BLeaf}^P(B)$ via the nondeterministic machine M. Computation paths of a nondeterministic machines can be followed in polynomial time if the nondeterministic choices are known. Defining A to consist of all pairs (x,p) such that p is a sequence of nondeterministic choices leading to a path of M that outputs "1" and $f(x)$ to be the number of paths of M on input x, we have $x \in L \iff \chi_{A_x}[0 \ldots f(x)] \in B$. $\qquad\square$

The definition of leaf languages allows for languages B not necessarily over the binary alphabet. If we want to come up with a connection to our generalized quantifier also for such B, we face a problem. In the definition in Sect. 2 the binary alphabet seems essential. Fortunately, for every B there is usually a $B' \subseteq \{0,1\}^*$ such that $\text{BLeaf}^P(B) = \text{BLeaf}^P(B')$, where B and B' are of the same complexity. In most cases, B' can simply be obtained from B by block encoding (then B and B' are FO-equivalent). We come back to this point in the next subsection.

3.2   The Complexity of a Leaf Language

In [HLS$^+$93] the question how complex a leaf language must be in order to characterize some given complexity class $\mathcal{K}$ was addressed. Let us start by considering some examples.

At great number of classes can be defined by regular leaf languages. This is obvious for NP, coNP and US as we saw in the previous section, for Mod_kP (all words with a number of "1"s divisible by k), but also true for higher levels of the polynomial hierarchy (see below) and the boolean hierarchy over NP (e.g. the class NP $\wedge$ coNP can be defined via the set of all words such that the string "010" appears at least once, but the string "0110" does not appear).

For other complexity classes, context-free languages come immediately to mind. PP can obviously be defined by the language maj from the previous section. Recalling the characterization of PSPACE via polynomial time alternating Turing machines, it is clear that the set of all (suitably encoded) boolean expressions involving the constants "true" and "false" and the connectives AND and OR that evaluate to "true", is an appropriate leaf language.

The question however arises if we can do better here. It was shown in [HLS$^+$93], that in the case of PSPACE there is a regular leaf language.

Let S_5 denote the word problem for the group of permutations on five elements (suitably encoded over the binary alphabet), i.e. S_5 consists of sequences of permutations which multiply out to the empty permutation.

Theorem 2. $(S_5)^{\text{PP}} = \text{PSPACE}$.

Proof sketch. For the inclusion from left to right, just observe that a PSPACE machine can traverse the whole computation tree of a given nondeterministic machine to evaluate the product over S_5. This simulation then stops accepting if and only if the result is the identity permutation.

For the other direction, we are given a language $L \in \text{PSPACE}$. Then there is a polynomial time alternating Turing machine accepting L. Thus, for every input w, machine M defines a polynomial depth computation tree $T(w)$ where the leafs carry values 0 or 1 and in the inner nodes the functions AND and OR are evaluated. $w \in L$ iff the root of this tree evaluates to 1. As a first step we transform this tree into a tree $T'(w)$ where in all the inner nodes the function NOR is evaluated. This can easily be achieved since the NOR function constitutes a complete basis for the boolean functions.

As a second step we now "simulate" NOR in S_5. This simulation is essentially due to David Barrington [Bar89]. Let b, c, d, e, f be the following permutations from the group S_5:

$$b = (23)(45), c = (12435), d = (243), e = (345), f = (152)$$

Further let a_0 be the empty permutation, denoted by $a_0 = ()$, and let $a_1 = (12345)$. Now consider the following product in S_5 including the variables x and y:

$$w(x, y) = a_0 b x^4 c y^4 d x e y f$$

Simple calculations show that $w(a_0, a_0) = a_1$ and $w(a_0, a_1) = w(a_1, a_0) = w(a_1, a_1) = a_0$. Thus coding the value *true* by a_1, and *false* by a_0, we can view w as the NOR-operation applied to x and y.

Now replace every appearance of a "NOR"-node in $T'(x)$ with sons x and y by a binary subtree of height 4 whose 16 leaves are

$$a_0 \ b \ x \ x \ x \ x \ c \ y \ y \ y \ y \ d \ x \ e \ y \ f$$

Thus we accept the input w if and only if the leaf string evaluates to a_1. Taking B to be the regular language

$$B =_{\text{def}} \{ \, x \mid x \text{ is a string of elements from } S_5 \text{ which evaluates to } a_1 \, \},$$

we then get $\text{Leaf}^P(B) = \text{PSPACE}$. It is easy to go from B to the word problem S_5 (by just adding one more factor a_1^{-1}), and since we have an identity element which we can insert arbitrarily in the leaf string to fill gaps in the computation tree in order to make it balanced, we get $\text{PSPACE} = \text{Leaf}^P(B) = \text{BLeaf}^P(S_5) = (S_5)^P P$.   □

The question now of course is what is so special about the language S_5. What can be said more generally? Using deep algebraic properties of regular languages exhibited in [Thé81,BT88] (see also the textbook [Str94]) one can show the following.

Let PH denote the union of all classes of the polynomial hierarchy [SM73], i.e. $\text{PH} = \text{NP} \cup \text{NP}^{\text{NP}} \cup \text{NP}^{\text{NP}^{\text{NP}}} \cup \cdots$. Let MOD-PH denote the oracle hierarchy constructed similarly, but now allowing as building blocks not only NP but also all classes $\text{Mod}_k P$ for arbitrary $k \in \mathbf{N}$.

Theorem 3 ([HLS⁺93]).

1. *Let A be a regular language whose syntactic monoid is non-solvable. Then $(A)^P P = \text{PSPACE}$.*
2. *Let SOLVABLE denote the class of all regular languages whose syntactic monoid is solvable. Then $(\text{SOLVABLE})^P P = \text{MOD-PH}$.*
3. *Let APERIODIC denote the class of all regular languages whose syntactic monoid is aperiodic. Then $(\text{APERIODIC})^P P = \text{PH}$.*

Regular leaf languages for individual levels of the polynomial hierarchy can also be given. For example Σ_2^p can be defined over $\Sigma = \{a, b, c\}$ by $\Sigma^* c a^+ c \Sigma^*$, intuitively: "there is a block consisting out of 'a's only". This is an $\exists \forall$ predicate directly reflecting the nature of Σ_2^p-computations. If we now chose a simple block encoding this might lead us out of the aperiodic languages. However, we may proceed as follows: Define $A_2 = (0+1)^*11(010)^+11(0+1)^*$. It is clear that this leaf language defines a subclass of Σ_2^p—just check that there are two substring 11 such that in between we have a sequence of occurrences of the 3-letter string 010; this is an $\exists \forall$ condition. On the other hand, suppose we are given a Σ_2^p machine M, i.e. an alternating machine with computation trees consisting of one level of

$\exists$ nodes followed by a second level of $\forall$ nodes; i.e. the initial configuration is the root of an existential tree where in the leaves we append universal subtrees. We transform this into a tree where we use the substring "11" in the leafstring as separator between different $\forall$ subtrees, and within each such subtree we simulate an accepting path by the 3 leaf symbols "010" and a rejecting path by the symbol "0". Then M produces a tree with at least one universal subtree consisting out of only accepting paths iff the leaf word of this simulation is in A_2. Σ_3^p can similarly be defined via $A_3 = (0+1)^*111\overline{A_2}111(0+1)^*$. This generalizes to higher levels of the polynomial hierarchy. With some care one can show that A_2 and A_3 are in levels $\mathcal{B}_2$ and $\mathcal{B}_3$, resp., of the Brzozowski-hierarchy of regular languages. This hierarchy of star-free regular languages measures the nesting depth of the dot (i.e. concatenation) operation. For a formal definition see [Eil76]. More generally the following holds:

Theorem 4 ([HLS$^+$93]). $(\mathcal{B}_k)^{\mathrm{P}}\mathrm{P}$ *is the boolean closure of the class* Σ_k^p.

Let us now come back to the question if PP (for which we gave a context-free leaf language above) can also be done by a regular language.

Corollary 1. PP *is not definable via a regular leaf language unless either* PP = PSPACE *or* PP $\subseteq$ MOD-PH.

Proof. If there is a regular leaf language L for PSPACE, then there are two cases to consider: either L is non-solvable (in this case PP = PSPACE) or L is solvable (then PP $\subseteq$ MOD-PH).

In [HLS$^+$93] leaf languages defined by restricting resource bounds as time and space were examined. It was shown that the complexity class obtained in this way is defined via the same resource, but the bound is one exponential level higher, for example $(\mathrm{P})^{\mathrm{P}}\mathrm{P} = \mathrm{EXPTIME}$, $(\mathrm{NP})^{\mathrm{P}}\mathrm{P} = \mathrm{NEXPTIME}$, $(\mathrm{LOGSPACE})^{\mathrm{P}}\mathrm{P} = \mathrm{PSPACE}$, $(\mathrm{PSPACE})^{\mathrm{P}}\mathrm{P} = \mathrm{EXPSPACE}$, and so on. Denoting the levels of the alternating log-time hierarchy [Sip83] by $\Sigma_k^{\log}$ ($k \in \mathbf{N}$), we get the following special case:

Theorem 5. $(\Sigma_k^{\log})^{\mathrm{P}}\mathrm{P} = \Sigma_k^p$.

3.3   Some Complexity Theoretic Applications

Normal Forms The characterization of PSPACE (Theorem 2) was somewhat surprising, since it points out a very restricted normal form for PSPACE computations. Cai and Furst defined a class $\mathcal{K}$ to be $\mathcal{K}'$-*serializable*, if every $\mathcal{K}$ computation can be organized into a number of local computations $c_1, \ldots, c_r$ (which in turn are restricted to be $\mathcal{K}'$ computations), each passing only a constant number k of bits as the result of its computation to the next local computation. The sequence $c_1, \ldots, c_r$ is uniform in the sense that there is one $\mathcal{K}'$ program that gets as input only the original input, a number i, and a string of k bits, and computes the k-bit-result of c_i's computation. Please refer to [CF91] for a formal definition. Machines as just described are also called bottleneck machines. The *bottleneck* refers to the restricted way of passing information onwards.

Corollary 2 ([HLS⁺93]). PSPACE *is* AC^0-*serializable.*

Proof sketch. Let $L \in BLeaf^P(S_5)$ via machine M. The information passed from one computation to the next will be an encoding of an element of the group S_5. Each local computation uses its number to recover from it a path of the nondeterministic Turing machine. (If the number does not encode a correct computation path, then we simply pass the information we get from our left neighbor onwards to the right.) The leaf symbol on this path is then multiplied to the permutation we got from the left, and the result is passed on to the right. This can be done in AC^0 since computation paths of polynomial time Turing machines can be checked in AC^0. (A computation path consists not only out of M's nondeterministic choices, but is a complete sequence of configurations of M.)

□

The power of bottleneck machines was examined in detail in [Her97]. He gave a connection between these machines and leaf languages defined via transformation monoids. The power of bottleneck machines as a function of the number of bits passed from one local computation to the next was determined.

Oracle Separations The original motivation for the introduction of leaf languages in [BCS92,Ver93] was the wish to have a uniform oracle separation theorem. Usually when relativized complexity classes are separated, this is achieved by constructing a suitable oracle by *diagonalization*, usually a stage construction. Bovet, Crescenzi, Silvestri, and Vereshchagin wanted to identify the common part of all these constructions in a unifying theorem, such that for future separations, one could concentrate more on the combinatorial questions which are often difficult enough. They showed that to separate two classes defined by leaf languages, it is sufficient to establish a certain non-reducibility between the defining languages. Let $A, B \subseteq \{0,1\}^*$. Say that A is polylogarithmic time bit-reducible to B, in symbols: $A \leq_m^{plt} B$, if there are two functions f, g computable in polylogarithmic time such that for all x, $x \in A \iff f(x,0)f(x,1)\cdots f(x,g(x)) \in B$.

Theorem 6 ([BCS92,Ver93]). *Let* $A, B \subseteq \{0,1\}^*$. *Then* $A \leq_m^{plt} B$ *if and only if for all oracles* Y, *the inclusion* $(A)^P P^Y \subseteq (B)^P P^Y$ *holds.*

Observe that $A \leq_m^{plt} B$ is just another formulation for the containment of A in $(B)^{plog}$POLYLOGTIME, which in turn is equivalent to the inclusion of the class $(A)^{plog}$POLYLOGTIME in $(B)^{plog}$POLYLOGTIME.

Corollary 3. *Let* $A, B \subseteq \{0,1\}^*$. *Then we have:*

$$(A)^{plog}\text{POLYLOGTIME} \subseteq (B)^{plog}\text{POLYLOGTIME}$$

if and only if for all oracles Y, *the inclusion*

$$(A)^P P^Y \subseteq (B)^P P^Y$$

holds.

In [BS97], Theorem 6 was strengthened as follows: It was shown that $(A)^P\mathcal{K} \subseteq (B)^P\mathcal{K}$ for all nontrivial classes $\mathcal{K}$ if and only if A is reducible to B by monotone polylogarithmic-time uniform projection reducibility. Refer to their paper for details.

Observe that a polylogarithmic time bit-reduction cannot (simply because of its time bound) read all of its input. This often allows one to prove $A \not\leq_m^{plt} B$ by an adversary arguments. We give a very simple example.

Example 1. Let $E = (0+1)^*1(0+1)^*$, $U = 1^*$ as in Sect. 2. Then $(E)^P\mathrm{P} = \mathrm{NP}$ and $(U)^P\mathrm{P} = \mathrm{coNP}$. Suppose $U \leq_m^{plt} E$. The input $x = 1^n$ must be mapped by this reduction to a word with at least one "1". The computation leading to this "1" however cannot read all of x. If we now define x' by complementing in x a bit which is not queried, then again x' will be mapped to a string in E, which is a contradiction. Thus $U \not\leq_m^{plt} E$, and hence there is an oracle separating coNP from NP.

Vereshchagin in [Ver93] used Theorem 6 to establish all relativizable inclusions between a number of prominent classes within PSPACE. His list contains besides the classes of the polynomial time hierarchy also UP, FewP, RP, BPP, AM, MA, PP, IP, and others.

A very satisfactory application of Theorem 6 was possible in the following special case. Say that $L \subseteq \Sigma^*$ is a *cardinal language*, if membership in L only depends on the frequency with which the elements of Σ appear in words. This means that if $\Sigma = \{a_1, \ldots, a_k\}$ we can associate L with a set $N(L) \subseteq \mathbf{N}^k$, in such a way that $w \in L$ iff there is a $(v_1, \ldots, v_k) \in N(L)$ where a_i occurs in w exactly v_i times $(1 \leq i \leq k)$. ($N(L)$ is the image of L under the Parikh mapping: $N(L) = \Psi_\Sigma(L)$.) Say that L is of *bounded significance* if there is a number $m \in \mathbf{N}$ such that for all $(v_1, \ldots, v_k)$ we have

$$(v_1, \ldots, v_k) \in N(L) \iff (\min(v_1, m), \ldots, \min(v_k, m)) \in N(L).$$

Using Ramsey theory, Hertrampf in [Her95a] proved the following:

Theorem 7 ([Her95a]). *There is an algorithm that, given two cardinal languages A, B of bounded significance, decides if $(A)^P\mathrm{P}^Y \subseteq (B)^P\mathrm{P}^Y$ for all oracles Y.*

Pushing his ideas just a bit further, the following was proved: We say that $p: \mathbf{N}^k \to \mathbf{N}$ is a positive linear combination of multinomial coefficients if $p(\boldsymbol{v}) = \sum_{u \leq z} \alpha_u \binom{v}{u}$ for some $z \in \mathbf{N}^k$, $\alpha_u \in \mathbf{N}$ (for $u \leq z$, the order taken componentwise).

Theorem 8 ([CHVW97]). *Let A, B be cardinal languages of bounded significance over a k element alphabet. Then $A \leq_m^{plt} B$ if and only if there are functions $p_1, \ldots, p_k: \mathbf{N}^k \to \mathbf{N}$ which are positive linear combinations of multinomial coefficients, such that for all $\boldsymbol{v} = (v_1, \ldots, v_k)$, $\boldsymbol{v} \in N(A)$ if and only if $(p_1(\boldsymbol{v}), \ldots, p_k(\boldsymbol{v})) \in N(B)$.*

In other words, if such k functions do not exist, then there is an oracle separating $(A)^P$P from $(B)^P$P. Thus we see that the oracle separation criterion Theorem 6 leads to a very strong statement in the context of cardinal languages. This result was used in [CHVW97] to establish a complete list of all relativizable inclusions between classes of the boolean hierarchy over NP and other classes defined by cardinal languages of bounded significance.

Valiant's counting class #P is of course strongly related to the notion of cardinal languages. In the case of #P we just deal with the binary alphabet, and we count the number of "1"s in a leaf string. Closure properties of #P, that is operations that don't lead us out of the class, play an important role to establish inclusions between complexity classes; e.g. Toda's result $PH \subseteq P^{PP}$ [Tod91] and Beigel, Reingold, and Spielman's proof that PP is closed under intersection [BRS91] both heavily build on the fact that #P is closed under certain sums, products, and choose operations.

Similar to Theorem 8 one can obtain the following:

Theorem 9 ([HVW95]). *A function $f : \mathbf{N}^k \to \mathbf{N}$ is a relativizable closure property of #P (i.e., relative to all oracles, if $h_1, \ldots, h_k \in$ #P then also $f(h_1, \ldots, h_k) \in$ #P), if and only if f is a positive linear combinations of multinomial coefficients.*

Circuit Lower Bounds Circuit classes as leaf languages have been considered in [CMTV96,Vol96a]. For background on circuit complexity, we refer the reader to [Str94]. It is immediate from Theorem 2 that $(NC^1)^P$P = PSPACE. Additionally one can prove e.g. that $(AC^0)^P$P = PH, and that $(TC^0)^P$P is the counting hierarchy CH, defined in [Wag86b,Wag86a] as $PP \cup PP^{PP} \cup PP^{PP^{PP}} \cup \cdots$. Finer results are given in [Vol97].

Building on leaf language characterizations, the circuit class TC^0 (where we require logtime uniformity) was separated from the counting hierarchy in [CMTV96]. This was improved by Allender [All96] to the following separation.

Theorem 10. $TC^0 \neq PP$.

Proof sketch. We sketch the proof of the weaker result from [CMTV96]. Suppose that $TC^0 = CH$. Then we have $TC^0 = CH = BLeaf^P(TC^0) = BLeaf^P(CH) \supseteq$ EXPTIME, thus $P \supseteq$ EXPTIME, which is a contradiction. Allender now observed that this can be extended to show that any language complete for PP under TC^0 reductions cannot be in TC^0.   □

In the non-uniform case no similar lower bound for TC^0 is known. If we relax the uniformity condition just a little bit, we know that

$$(\text{logspace-uniform } AC^0)^P\text{P} = \text{PSPACE}$$

(thus also $BLeaf^P(\text{logspace-uniform } TC^0) = \text{PSPACE}$). This shows that logtime-uniformity is critical in the above proof.

In Corollary 3 it became clear that the oracle separability of two polynomial time classes is equivalent to the absolute separability of two lower classes with the same acceptance paradigm. A similar relation is known between polynomial time and constant depth circuit classes. E.g. building on previous work by Furst, Saxe, and Sipser [FSS84], Yao in his famous paper used a lower bound for the parity function to construct an oracle separating PSPACE from the polynomial hierarchy [Yao85]. This connection has been exploited a number of times since then.

The formal connection between Theorem 6 and the Furst, Saxe, Sipser approach to oracle construction has been given in [Vol97]. The main observation that has to be made is that $\leq_m^{plt}$-reductions can be performed by (uniform) qAC^0 circuits. qAC^0 stands for quasipolynomial AC^0 [Bar92], i.e. unbounded fan-in circuits of constant depth and size $2^{\log^{O(1)} n}$. (Similarly we will also use qTC^0 for quasipolynomial size TC^0 circuits, and qNC^1 for quasipolynomial size NC^1 circuits.)

Theorem 11. *Let $A, B \subseteq \{0,1\}^*$. Then we have: $A \notin (B)^{\text{plog}}\text{qAC}^0$ if and only if $(A)^{\text{plog}}\text{qAC}^0 \not\subseteq (B)^{\text{plog}}\text{qAC}^0$ if and only if there is an oracle Y such that $(A)^{\text{P}}\text{PH}^Y \not\subseteq (B)^{\text{P}}\text{PH}$.*

This theorem can be used to attack the "nagging question" [For97] how to separate superclasses of P^{PP} from PSPACE. Some special cases are the following.

Corollary 4. *$S_5 \notin \text{qTC}^0$ if and only if $\text{qTC}^0 \neq \text{qNC}^1$ if and only if there is an oracle separating the counting hierarchy from PSPACE.*

Proof sketch. Under the assumption $S_5 \in \text{qTC}^0$, the following inclusion chain holds relativizably:

$$\text{PSPACE} = \text{BLeaf}^{\text{P}}(S_5) \subseteq \text{BLeaf}^{\text{P}}(\text{qTC}^0) = \text{CH}.$$

This proves the direction from right to left. For the other direction, if relative to all oracles PSPACE $\subseteq$ CH then S_5 polylogarithmic time bit-reduces to qTC^0, but this class is even closed under qAC^0 reductions.   □

Define par $=_{\text{def}} \{ w \in \{0,1\}^* \mid$ the number of "1"s in w is odd $\}$, and let maj be as in Sect. 2.

Corollary 5. *$S_5 \notin (\text{maj})^{\text{plog}}(\text{par})^{\text{plog}}\text{qAC}^0$ if and only if there is an oracle separating $\text{PP}^{\oplus\text{P}}$ from PSPACE.*

Proof sketch. If PSPACE $\subseteq \text{PP}^{\oplus\text{P}}$ then S_5 polylogarithmic time bit-reduces to a language in the class $(\text{maj})^{\text{plog}}(\text{par})^{\text{plog}}\text{qAC}^0$, and therefore S_5 is even in this class (it is closed under $\leq_m^{plt}$).

On the other hand, if $S_5 \in (\text{maj})^{\text{plog}}(\text{par})^{\text{plog}}\text{qAC}^0$, then PSPACE $=$ $\text{BLeaf}^{\text{P}}(S_5) \subseteq \text{BLeaf}^{\text{P}}((\text{maj})^{\text{plog}}(\text{par})^{\text{plog}}\text{qAC}^0) = \text{PP}^{\oplus\text{P}^{\text{PH}}} = \text{PP}^{\oplus\text{P}}$.   □

A refinement of Theorem 11 and further investigations along these lines can be found in [Vol97].

3.4   Definability vs. Tree Shapes

Our quantifier from Sect. 2 coincides as we saw in the polynomial time context with leaf languages for balanced computation trees. The unbalanced case has also attracted some attention in the literature. It was observed in [HVW96] that the relativization result from [BCS92,Ver93] *does not hold* in the case of unbalanced trees. Thus, part of the motivation to consider this construct is gone. Nevertheless definability questions are also interesting in this case. The just mentioned observation even makes a systematic comparison of both models a worthwhile study.

Balanced vs. Unbalanced Trees In [HVW96] the question of definability of the polynomial hierarchy was addressed. As mentioned earlier in Theorem 5, the classes of the log-time hierarchy exactly define the classes of the polynomial hierarchy. However, in the case of unbalanced trees, one can somehow use the tree structure to hide an oracle that is able to count paths. More formally,

Theorem 12 ([HVW96]). $\mathrm{Leaf}^{\mathrm{P}}(\Sigma_k^{\log}) = (\Sigma_k^{\mathrm{p}})^{\mathrm{PP}}$.

The Acceptance Power of Different Tree Shapes Hertrampf [Her95b] considered besides the above two models also the definition of classes via leaf languages for computation trees which are *full binary trees*. The obtained classes are noted by $\mathrm{FBTLeaf}^{\mathrm{P}}(\cdot)$. Though trivially for every $B \subseteq \{0,1\}^*$ we have $\mathrm{FBTLeaf}^{\mathrm{P}}(B) \subseteq \mathrm{BLeaf}^{\mathrm{P}}(B) \subseteq \mathrm{Leaf}^{\mathrm{P}}(B)$, Hertrampf proved the somewhat counterintuitive result, that the definability power by arbitrary single regular languages does not decrease but possibly *increases* as the tree shapes get more and more regular; that is for every regular language B there is a regular language B' such that $\mathrm{Leaf}^{\mathrm{P}}(B) = \mathrm{Leaf}^{\mathrm{P}}(B') = \mathrm{BLeaf}^{\mathrm{P}}(B')$, and for every regular language B there is a regular language B' such that $\mathrm{BLeaf}^{\mathrm{P}}(B) = \mathrm{BLeaf}^{\mathrm{P}}(B') = \mathrm{FBTLeaf}^{\mathrm{P}}(B')$.

Definability Gaps In the case of arbitrary tree shapes, Borchert et al. were able to prove the existence of definability gaps. In particular, the following was shown.

Theorem 13. *Suppose the polynomial hierarchy does not collapse, and let B be an arbitrary regular language.*

1. *If* $\mathrm{P} \subseteq \mathrm{Leaf}^{\mathrm{P}}(B) \subseteq \mathcal{K}$, *then* $\mathrm{Leaf}^{\mathrm{P}}(B) = \mathrm{P}$ *or* $\mathrm{Leaf}^{\mathrm{P}}(B) = \mathcal{K}$, *where* $\mathcal{K}$ *is one of the classes* NP, coNP, *or* $\mathrm{Mod}_p\mathrm{P}$ *(for some prime number p)* [Bor94].
2. *If* NP $\subseteq \mathrm{Leaf}^{\mathrm{P}}(B) \subseteq$ coUS, *then* $\mathrm{Leaf}^{\mathrm{P}}(B) =$ NP *or* $\mathrm{Leaf}^{\mathrm{P}}(B) =$ coUS [BKS96] *(analogously for* coNP *and* US).

We come back to questions of this kind in Sect. 6.

4   Other Resource Bounds

4.1   Circuit Classes

Corollary 2 easily yields the following:

Corollary 6. $(S_5)^P AC^0 = PSPACE$.

This coincidence between $(\cdot)^P P$ and $(\cdot)^P AC^0$ holds under more general circumstances. Let $\mathcal{N}$ denote the set of all languages $L \subseteq \Sigma^*$ that contain a neutral letter e, i.e. for all $u, v \in \Sigma^*$, we have $uv \in L \iff uev \in L$.

Theorem 14. *If* $B \in \mathcal{N}$ *then* $(B)^P P = (B)^P AC^0$.

Proof sketch. Computation paths of nondeterministic Turing machines can be checked in AC^0 as already pointed out in the proof of Corollary 2. The required AC^0 computation in input (x, y) now checks that its second input argument is a correct path of the corresponding machine on input x; if so it outputs 1 iff this path is accepting and 0 otherwise. If y does not encode a correct path then the neutral letter is output.   □

A careful inspection of the just given proof reveals that the result not only holds for language $B \in \mathcal{N}$, $B \subseteq \{0, 1\}^*$, but also for languages B that are obtained from some $B' \in \mathcal{N}$, $B \subseteq \Sigma^*$ (possibly $|\Sigma| > 2$) by block encoding. The same generalization holds for all results that we state below for "$B \in \mathcal{N}$" (i.e., Theorem 20 and all results in Sect. 5).

In the context of NC^1 and subclasses, some interesting results can be obtained for classes of the form $(\cdot)^{\log} AC^0$.

First, Barrington's theorem [Bar89] yields:

Theorem 15. $(S_5)^{\log} AC^0 = NC^1$.

Theorem 16.   *1.* $(B)^{\log} AC^0 = NC^1$ *for every regular language* B *whose syntactic monoid is non-solvable.*
   2. $(SOLVABLE)^{\log} AC^0 = ACC^0$.

Generally the class $(B)^{\log} AC^0$ roughly corresponds to AC^0 circuits with a B gate on top, e.g. $(maj)^{\log} AC^0$ is the class of all languages accepted by perceptrons.

AC^0 circuits with arbitrary B gates are examined in [BIS90,BI94] (see also Sect. 6).

4.2   Logspace and Logtime Leaf Languages

In the same spirit as above for nondeterministic *polynomial time* machines, Jenner, McKenzie, and Thérien examined in [JMT94] leaf languages for nondeterministic *logarithmic time* and *logarithmic space* machines.

First turning to the logspace case, we observe that the trivial way to formulate $\mathrm{Leaf}^{\mathrm{L}}(B)$, the class defined by logspace machines with leaf language B, as a class $(\cdot)^{\mathrm{P}}\mathrm{L}$ does not work (L denotes the class of logspace decidable sets). This is because (for $B \in \mathcal{N}$) already $(B)^{\mathrm{p}}\mathrm{P} = (B)^{\mathrm{p}}\mathrm{AC}^0$ (see Sect. 4.1), and therefore also $(B)^{\mathrm{p}}\mathrm{P} = (B)^{\mathrm{p}}\mathrm{L}$.

However, if we turn to logarithmic space-bounded one-way protocol machines or 2-1-machines [Lan86], we can come up with a connection. A *2-1-Turing machine* is a Turing machine with two input tapes: first a (regular) input tape that can be read as often as necessary, and second, an additional (protocol) tape that can be read only once (from left to right). Define 2-1-L to be the class of all two argument languages L that can be computed by logspace-bounded 2-1-TMs such that in the initial configuration, the first argument of the input is on the regular input tape, and the second argument is on the one-way input tape. Then the following can be shown using ideas from [Lan86]:

Theorem 17. *Let $B \subseteq \{0,1\}^*$. Then $(B)^{\mathrm{p}}$2-1-L $= \mathrm{Leaf}^{\mathrm{L}}(B)$.*

Jenner, McKenzie, and Thérien showed that in a lot of cases, the balanced and unbalanced model coincide for logarithmic space machines, and moreover it sometimes coincides with the polynomial time case, e.g. Theorem 5 above also holds with leaf languages for logspace machines. Interesting to mention is that in the logarithmic space model, regular leaf languages define the class P, while NC^1 defines the class PSPACE.

In the logarithmic time case, coincidence with the logarithmic time reducibility closure could be shown for all well-behaved leaf languages. Formulated in terms of our quantifier, some of their results read as follows:

Theorem 18 ([JMT94]).

1. $(\mathrm{REG})^{\log}\mathrm{DLOGTIME} = \mathrm{NC}^1$.
2. $(\mathrm{CFL})^{\log}\mathrm{DLOGTIME} = \mathrm{LOGCFL}$.
3. $(\mathrm{CSL})^{\log}\mathrm{DLOGTIME} = \mathrm{PSPACE}$.

Theorem 19 ([JMT94]).

1. $(B)^{\log}\mathrm{DLOGTIME} = \mathrm{NC}^1$ *for any regular language B whose syntactic monoid is non-solvable.*
2. $(\mathrm{SOLVABLE})^{\log}\mathrm{DLOGTIME} = \mathrm{ACC}^0$.
3. $(\mathrm{APERIODIC})^{\log}\mathrm{DLOGTIME} = \mathrm{AC}^0$.

4.3 Other Models

Type 2 Operators Operators ranging not over words but over oracles, so called *type 2 operators*, have been examined in [BW96,BVW96,VW97] and elsewhere. Most of the considered classes coincide with classes of the form $(B)^{\mathcal{F}}\mathcal{K}$ where $\mathcal{K} = \text{coNP}$ or $\mathcal{K} = \text{PSPACE}$ and $\mathcal{F}$ is the class of all exponential time computable functions (let us write $(B)^{\exp}\mathcal{K}$ as a shorthand for this choice of $\mathcal{F}$). A word of care about the computational model however is in order now. We say that a language L belongs to the class $(B)^{\exp}\text{coNP}$ if there is a function f computable in exponential time, and a set A such that $x \in L \iff \chi_{A_x}[0 \ldots f(x)] \in B$, where A is accepted by some co-nondeterministic Turing machine M that on input $\langle x, y \rangle$ runs in time polynomial in the length of x. The length of y is possibly exponential in the length of x; thus to enable M to access all positions of y within its time bound we supply M with a regular input tape on which x is found, and a second input tape for y, which is accessed by an index tape . This special input tape is similar to an oracle tape, and therefore quantifiers over strings on this tape translate to quantifiers over oracles. (In the case of $(B)^{\exp}\text{PSPACE}$ we require our machines to use space no more than polynomial in the length of their regular input x.)

Theorem 20. *Let $B \in \mathcal{N}$. Then we have:*

$$(B)^{\exp}\text{EXPTIME} = (B)^{\exp}\text{PSPACE} = (B)^{\exp}\text{coNP}.$$

Proof sketch. If we look at the proof of Theorem 14 we see that to check correct computation paths we actually don't need the full power of AC^0. $\Pi_1^{\log}$ is sufficient, but we have to modify the computation model slightly as follows: The log-time machine has a regular input tape (which is accessed as usual by using an index tape) and a second input tape on which the path to be checked is given (again access is by an index tape). We thus get:

$$(B)^{\text{P}}\text{P} = (B)^{\text{P}}\text{AC}^0 = (B)^{\text{P}}\Pi_1^{\log}.$$

Using standard translation arguments we now get the claim of the theorem by lifting up this equation one exponential level.   □

NC^1 Leaf Languages In [CMTV96] leaf languages for nondeterministic finite automata were considered. The original input is however first given into a uniform projection, and the result of this projection is then fed into the NFA. Barrington's Theorem 15 implies that with regular leaf languages we thus get exactly the class NC^1. Some other characterizations were given in [CMTV96], and the model was also used to examine counting classes within NC^1.

Function Classes In [KSV97] the definability of function classes has been examined. An oracle separation criterion generalizing Theorem 6 was given and applied successfully in some open cases.

5   Leaf Languages vs. Lindström Quantifiers

Lindström quantifiers [Lin66] are a well established generalized quantifier notion in finite model theory. The reader probably has noticed some resemblance of our definition in Sect. 2 with that of Lindström quantifiers. It will be our aim in the upcoming sections to make this precise.

As we will see there is a strong connection between leaf languages for polynomial time machines and *second-order* Lindström quantifiers. Since this notion might not be so well-known, we give—after very briefly recalling some terminology from finite model theory—a precise definition in Sect. 5.1.

In later subsections we will have the need to talk about *the second-order version of a given first-order Lindström quantifier*. We chose to make this precise by talking about the semantics of quantifiers given by *languages* instead of the usual way of defining semantics by classes of structures. In the next subsection, we will define how a language B gives rise to a first-order quantifier Q_B^0 and a second order quantifier Q_B^1.

5.1   Second-Order Lindström Quantifiers

A signature is a finite sequence $\tau = \langle R_1, \ldots, R_k, c_1, \ldots, c_\ell \rangle$ of relation symbols and constant symbols. A finite structure of signature τ is a tuple $\mathcal{A} = (A, R_1^{\mathcal{A}}, \ldots, R_k^{\mathcal{A}}, c_1^{\mathcal{A}}, \ldots, c_\ell^{\mathcal{A}})$ consisting of a finite set A (the universe of $\mathcal{A}$) and interpretations of the symbols in τ by relations over $\mathcal{A}$ (of appropriate arity) and elements of $\mathcal{A}$. Struct(τ) is the set of all *finite ordered structures* over τ. The *characteristic string* χ_R of a relation $R \in \{0, \ldots, n-1\}^a$ is the string $\chi_R =_{\text{def}} b_1 \cdots b_{n^a}$ where $b_i = 1$ iff the i-th vector in $\{0, \ldots, n-1\}^a$ (in the order $(0, \ldots, 0, 0) < (0, \ldots, 0, 1) < (n-1, \ldots, n-1, n-1)$) is in R. For $1 \leq i \leq n^a$, let $\chi_R[i]$ denote the i-th bit in χ_R.

If $\mathcal{L}$ is a logic (as e.g. FO or SO) and $\mathcal{K}$ is a complexity class, then we say that $\mathcal{L}$ *captures* $\mathcal{K}$ if every property over (standard encodings of) structures decidable within $\mathcal{K}$ is expressible by $\mathcal{L}$ sentences, and on the other hand for every fixed $\mathcal{L}$ sentence ϕ, determining whether $\mathcal{A} \models \phi$ can be done in $\mathcal{K}$. As an abbreviation we will most of the time simply write $\mathcal{K} = \mathcal{L}$.

A first-order formula ϕ with k free variables defines for every structure $\mathcal{A}$ the relation $\phi^{\mathcal{A}} =_{\text{def}} \{ a \in A^k \mid \mathcal{A} \models \phi(a) \}$, see [EF95].

Every class of structures $K \subseteq \text{Struct}(\sigma)$ over a signature $\sigma = \langle P_1, \ldots, P_s \rangle$ defines the first-order Lindström quantifier Q_K as follows: Let $\phi_1, \ldots, \phi_s$ be first-order formulae over signature τ such that for $1 \leq i \leq s$ the number of free variables in ϕ_i is equal to the arity of P_i. Then

$$Q_K x_1, \ldots, x_s \, [\phi_1(x_1), \ldots, \phi_s(x_s)]$$

is a Q_KFO formula. If $\mathcal{A} \in \text{Struct}(\tau)$, then

$$\mathcal{A} \models Q_K x_1, \ldots, x_s \, [\phi_1(x_1), \ldots, \phi_s(x_s)]$$

iff $(A, \phi_1^{\mathcal{A}}, \ldots, \phi_s^{\mathcal{A}}) \in K$.

The just given definition is the original definition given by Lindström [Lin66], which the reader will also find in textbooks, see e.g. [Ebb85,EF95]. For our examinations, the following equivalent formulation will be useful (observe that this only makes sense for ordered structures):

Given a first-order formula ϕ with k free variables and a corresponding finite ordered structure $\mathcal{A}$, this defines the binary string $\chi_{\phi^{\mathcal{A}}}$ of length n^k ($n = |A|$). Now given a sequence $\phi_1, \ldots, \phi_s$ of formulae with k free variables each and a structure $\mathcal{A}$, we similarly get the tuple $(\chi_{\phi_1^{\mathcal{A}}}, \ldots, \chi_{\phi_s^{\mathcal{A}}})$, where $|\chi_{\phi_1^{\mathcal{A}}}| = \cdots = |\chi_{\phi_s^{\mathcal{A}}}| = n^k$. Certainly, there is a one-one correspondence between such tuples and strings of length n^k over a larger alphabet (in our case with 2^s elements) as follows. Let A_s be such an alphabet. Fix an arbitrary enumeration of A_s, i.e. $A_s = \{a_0, a_1, \ldots, a_{2^s-1}\}$. Then $(\chi_{\phi_1^{\mathcal{A}}}, \ldots, \chi_{\phi_s^{\mathcal{A}}})$ corresponds to the string $b_1 b_2 \cdots b_{n^k}$, where for $1 \le i \le n^k$, $b_i \in A_s$, $b_i = a_k$ for that k whose length s binary representation (possibly with leading zeroes) is given by $\chi_{\phi_1^{\mathcal{A}}}[i] \cdots \chi_{\phi_s^{\mathcal{A}}}[i]$. In symbols: $w_s(\chi_{\phi_1^{\mathcal{A}}}, \ldots, \chi_{\phi_s^{\mathcal{A}}}) = b_1 b_2 \cdots b_{n^k}$.

This leads us to the following definition: A sequence $[\phi_1, \ldots, \phi_s]$ is in *first-order word normal form*, iff the ϕ_i have the same number k of free variables. Let Γ be an alphabet such that $|\Gamma| \ge 2^s$, and let $B \subseteq \Gamma^*$. Then $\mathcal{A} \models Q_B \boldsymbol{x} [\phi_1(\boldsymbol{x}), \ldots, \phi_s(\boldsymbol{x})]$ iff $w_s(\chi_{\phi_1^{\mathcal{A}}}, \ldots, \chi_{\phi_s^{\mathcal{A}}}) \in B$.

It can be shown [BV96,Bur96] that every Lindström quantifier Q_K can without loss of generality be assumed to be of the form Q_B as just defined. This is the case since for every sequence $[\phi_1, \ldots, \phi_s]$ of first-order formulae we find an equivalent sequence in word normal form such that the corresponding formulae with Lindström quantifier express the same property.

Second-order Lindström quantifiers are defined as follows [BV96,Bur96]: Given a formula ϕ with free second-order variables $P_1, \ldots, P_m$ and a structure $\mathcal{A}$, define $\phi^{2^{\mathcal{A}}} =_{\text{def}} \{ (R_1^{\mathcal{A}}, \ldots, R_m^{\mathcal{A}}) \mid \mathcal{A} \models \phi(R_1^{\mathcal{A}}, \ldots, R_m^{\mathcal{A}}) \}$, and let $\chi_{\phi^{2^{\mathcal{A}}}}$ be the corresponding characteristic string, the order of vectors of relations being the natural one induced by the underlying order of the universe. If the arities of $P_1, \ldots, P_m$ are $r_1, \ldots, r_m$, resp., then the length of $\chi_{\phi^{2^{\mathcal{A}}}}$ is $2^{n^{r_1} + \cdots + n^{r_m}}$

Let $\sigma = \langle \sigma_1, \ldots, \sigma_s \rangle$ be a signature, where $\sigma_i = \langle P_{i,1}, \ldots, P_{i,m_i} \rangle$ for $1 \le i \le s$. Thus σ is a signature consisting of a sequence of s signatures with only predicate symbols each. Let $\ell_{i,j}$ be the arity of $P_{i,j}$. A *second-order structure* of signature σ is a tuple $\mathcal{A} = (A, \mathcal{R}_1, \ldots, \mathcal{R}_s)$, where for every $1 \le i \le s$, $\mathcal{R}_i \subseteq \{ (R_{i,1}, \ldots, R_{i,m_i}) \mid R_{i,j} \subseteq A^{\ell_{i,j}} \}$. Given now a signature τ and second-order formulae $\phi_1(\boldsymbol{X}_1), \ldots, \phi_s(\boldsymbol{X}_s)$ over τ where for every $1 \le i \le s$ the number and arity of free predicates in ϕ_i corresponds to σ_i. Let $\mathcal{K}$ be a class of second-order structures over σ. Then $Q_{\mathcal{K}} \boldsymbol{X}_1, \ldots, \boldsymbol{X}_s [\phi_1(\boldsymbol{X}_1), \ldots, \phi_s(\boldsymbol{X}_s)]$ is a $Q_{\mathcal{K}}$SO formula. If $\mathcal{A} \in \text{Struct}(\tau)$, then $\mathcal{A} \models Q_{\mathcal{K}} \boldsymbol{X}_1, \ldots, \boldsymbol{X}_s [\phi_1(\boldsymbol{X}_1), \ldots, \phi_s(\boldsymbol{X}_s)]$ iff $(A, \phi_1^{2^{\mathcal{A}}}, \ldots, \phi_s^{2^{\mathcal{A}}}) \in \mathcal{K}$.

Again, we want to talk about second-order Lindström quantifiers defined by languages. Thus we define analogously to the above: A sequence $[\phi_1(\boldsymbol{X}_1), \ldots, \phi_s(\boldsymbol{X}_s)]$ of second-order formulae is in *second-order word normal form*, if the $\phi_1, \ldots, \phi_s$ have the same predicate symbols, i.e. in the above terminology $\sigma_1 =$

$\cdots = \sigma_s = \langle P_1, \ldots, P_m \rangle$. Let for $1 \leq i \leq m$ the arity of P_i be r_i. Observe that in this case, $|\chi_{\phi_1^{2^A}}| = \cdots = |\chi_{\phi_s^{2^A}}| = 2^{n^{r_1} + \cdots + n^{r_m}}$ (for $n = |A|$), thus $(\chi_{\phi_1^{2^A}}, \ldots, \chi_{\phi_s^{2^A}})$ corresponds to a word of the same length over an alphabet of cardinality 2^s. Given now a language $B \subseteq \Gamma^*$ with $|\Gamma| \geq 2^s$, the second-order Lindström quantifier given by B is defined by $\mathcal{A} \models Q_B^1 \boldsymbol{X} [\phi_1(\boldsymbol{X}), \ldots, \phi_s(\boldsymbol{X})]$ iff $w_s(\chi_{\phi_1^{2^A}}, \ldots, \chi_{\phi_s^{2^A}}) \in B$.

Again it was shown in [BV96,Bur96] that for every second-order Lindström quantifier $Q_\mathcal{K}$ there is an equivalent Q_B^1.

When talking about the first-order Lindström quantifier given by B, we sometimes explicitly write Q_B^0 instead of Q_B. In addition to the above logics $Q_B^0 \mathrm{FO}$ and $Q_B^1 \mathrm{SO}$ where we allow Lindström quantifiers followed by an arbitrary first-order (second-order, resp.) formula, we also need $Q_B^1 \mathrm{FO}$ (where we have a second-order Lindström quantifier followed by a formula with no other second-order quantifiers), and $\mathrm{FO}(Q_B^0)$ and $\mathrm{SO}(Q_B^1)$ (where we have first-order (second-order, resp.) formulae with arbitrary nesting of universal, existential, and Lindström quantifiers). For a class of languages $\mathcal{C}$ we use the notation $Q_\mathcal{C}$ with the obvious meaning, e.g. $\mathrm{FO}(Q_\mathcal{C}^0)$ denotes all first-order sentences with arbitrary quantifiers Q_B^0 for $B \in \mathcal{C}$.

5.2   A Logical Characterization of the Leaf Concept

The main technical connection between polynomial time leaf languages and second-order Lindström quantifiers is given in the following theorem:

Theorem 21. *Let M be a polynomial time nondeterministic machine whose computation tree is always a full binary tree, and let $B \subseteq \{0,1\}^*$. Then there is a Σ_1^1 formula ϕ such that*

$$\mathrm{Leaf}^M(B) = Q_B^1 \boldsymbol{X} [\phi(\boldsymbol{X})].$$

Proof sketch. We use a modification of Fagin's proof [Fag74]. The Q_B^1 quantifier will bind the nondeterministic guesses of the machine. The second-order quantifiers in ϕ will bind variables Y that encode computation paths of M. The formula $\phi(X)$ says "there is a Y encoding a correct computation path of M corresponding to nondeterministic guesses X, which is accepting." $\square$

If we deal with $B \subseteq \Gamma^*$ not necessarily over the binary alphabet, then instead of ϕ above, we get formulae $\phi_s, \ldots, \phi_s$ such that

$$\mathrm{Leaf}^M(B) = Q_B^1 \boldsymbol{X} [\phi_1(\boldsymbol{X}), \ldots, \phi_s(\boldsymbol{X})].$$

$\phi_i(X)$ says "there is a Y encoding a correct computation path of M corresponding to nondeterministic guesses X, and the leaf symbol produced on this path has a 1 in bit position i (in binary). Thus what we have here is some block-encoding of Γ in binary strings of length s.

The just given theorem shows that $\mathrm{FBTLeaf}^\mathrm{P}(B) \subseteq Q_B^1 \Sigma_1^1$. The question now of course is if there is a logic capturing $\mathrm{FBTLeaf}^\mathrm{P}(B)$. For the special case $B \in \mathcal{N}$, the answer is yes.

Theorem 22 ([BV96]). *Let $B \in \mathcal{N}$. Then $Q_B^1 \mathrm{FO} = \mathrm{BLeaf}^P(B)$.*

Proof sketch. This time Q_B^1 binds the nondeterministic guesses X as well as the encoding Y of a possible computation path. The first order formulae "output" the neutral letter, if Y does not encode a correct path. This proves the direction from right to left. For the other inclusion, we observe that we can design a Turing machine which branches on all possible assignments for the relational variables and then simply evaluates the first-order part. □

In the preceding theorem $\mathrm{BLeaf}^P(B)$ is captured by the logic $Q_B^1 \mathrm{FO}$ *uniformly* in the sense of [MP93,MP94]; this means that the particular formula describing the Turing machine is independent of the leaf language.

Let us next address the question if the quantifier in the preceding theorem is genuinely second-order. First, we have to give some definitions. A succinct representation [Wag86a,BLT92,Vei96] of a binary word x is a boolean circuit giving on input i the ith bit of x. The succinct version sA of a language A is the following: Given a boolean circuit describing a word x, is $x = x_1 0 x_2 0 \cdots x_{n-1} 0 x_n 1 w$ for arbitrary $w \in \{0,1\}^*$, such that $x_1 x_2 \cdots x_n \in A$? The boolean circuits we allow are standard unbounded fan-in circuits over AND, OR, NOT. The encoding consists of a sequence of tuples (g, t, h), where g and h are gates, t is the type of g, and h is an input gate to g (if g is not already an input variable).

Now we see that there is an equivalent first-order logic for $Q_B^1 \mathrm{FO}$.

Theorem 23. *Let $B \in \mathcal{N}$. Then $\mathrm{BLeaf}^P(B) = Q_B^1 \mathrm{FO} = Q_{sB}^0 \mathrm{FO}$.*

Proof sketch. Veith [Vei96] showed that sB is complete for $\mathrm{BLeaf}^P(B)$ under projection reductions. (A somewhat weaker result appeared in [BL96]). This together with Theorem 22 implies the theorem. □

5.3 Applications

Burtschick and Vollmer in [BV96] also examined logically defined leaf languages. It turned out that if the leaf language is given by a first-order formula, then the obtained complexity class is captured by the corresponding second-order logic. More specifically, they proved for instance:

Theorem 24 ([BV96]). *Let $B \in \mathcal{N}$. Then $(Q_B^0 \Sigma_k^0)^P \mathrm{P} = Q_B^1 \Sigma_k^1$.*

As a special case of Theorem 24 we get a characterization of the classes of the polynomial hierarchy which is tighter than the one in Theorem 5.

Corollary 7. $(\Sigma_k^0)^P \mathrm{P} = \Sigma_k^p$.

¿From the PSPACE characterization Theorem 2 and the above results, we get the following model-theoretic characterization of PSPACE:

Corollary 8. $Q_{S_5}^1 \mathrm{FO} = Q_{sS_5}^0 \mathrm{FO} = \mathrm{PSPACE}$.

5.4   First-Order Quantifiers

It is known from the work of Immerman et al. [Imm89,BIS90] that (uniform) AC^0 is captured by FO. However, for this result, we have to include the bit predicate in our logic. We make this assumption throughout this subsection (all the previously given results are valid without the bit predicate).

Theorem 25. *Let $B \subseteq \{0,1\}^*$. Then $(B)^{\log} AC^0 = Q_B^0 FO$.*

Theorem 25, together with results from Sect. 4.2 on logtime leaf languages, gives some more model-theoretic characterizations.

Corollary 9.   *1. PSPACE $= Q_{CSL}^0 FO = FO(Q_{CSL})$.*
2. LOGCFL $= Q_{CFL}^0 FO = FO(Q_{CFL})$.

Proof sketch. One can show that generally $\text{Leaf}^{LT}(B) \subseteq Q_B^0 FO$. The corollary then follows from Theorem 18. $\square$

6   Conclusion

We examined a generalized quantifier notion in computational complexity. We proved that not only all quantifiers examined so far (whether in the logarithmic, polynomial, or exponential time context) can be seen as special cases of this quantifier, but also circuits with generalized gates and Turing machines with leaf language acceptance.

Most of the emerging complexity classes can be characterized by means from finite model theory. We gave a precise connection to finite model theory by showing how complexity classes defined by the generalized quantifier relate to classes of finite models defined by logics enhanced with Lindström quantifiers.

A number of questions remain open. The results we gave in Sect. 5 related complexity classes to logics of the form "Lindström quantifier followed by a usual first- or second-order formula." It is not clear if logics defined by arbitrary nesting of Lindström quantifiers have a nice equivalent in terms of the generalized complexity theoretic quantifier. Barrington, Immerman, and Straubing proved:

Theorem 26 ([BIS90]). *Let $B \in \mathcal{N}$. Then $FO(Q_B^0) = AC^0[B]$ (AC^0 circuits with B gates).*

Moreover one can show:

Theorem 27 ([Vol96b]). *Let $B \in \mathcal{N}$. Then $FO(Q_B^1)$ is the oracle hierarchy given by $(B)^P AC^0$ as building block.*

But the general relationship remains unclear. The work of Makowsky and Pnueli (see [MP93,MP94]), Stewart (see e.g. [Ste91,Ste92]), and Gottlob (see [Got95]) shows that there is a strong relation between Lindström logics and relativized computation. The just mentioned results also hint in that direction. Gottlob [Got95] related the expressive power of logics of the form "Lindström quantifier Q followed by first-order formula" to the expressive power of $FO(Q)$. However his results only apply for superclasses of L (logarithmic space). Interesting cases within NC^1 remain open. Generally the connection between prenex Lindström logics vs. logics allowing arbitrary quantifier nestings on the model theoretic side, and leaf languages vs. oracle computations on the complexity theoretic side should be made clearer.

It is open for which of the results in Sect. 5.4 the bit predicate is really needed. One can show that without bit, $Q_{CFL}FO = CFL$ contrasting the corresponding result with bit given in Corollary 9. The power of the bit predicate in this context deserves further attention.

From a complexity theoretic point of view, we think the main open question is the following. A lot of classes defined by leaf languages have been identified. However, most of the results are not about singular leaf languages but about classes of leaf languages. For example (see Theorem 3), if we take an arbitrary aperiodic leaf languages, then the complexity class we obtain is included in PH, and conversely we get all of PH when we allow aperiodic leaf languages: $BLeaf^P(APERIODIC) = PH$. The question now is the following: What exactly are the classes of the form $BLeaf^P(B)$ for aperiodic B? Is it possible to come up with a complete list of classes that can be defined in this way? Some of the results in Sect. 3.4 point in this direction. For example we know that there is no class between P and NP that can be defined by a regular leaf language (unfortunately the result given in Sect. 3.4 holds only for the unbalanced case). Can we come up with similar result for the balanced case? Generally, very little is known about the power of *single* leaf languages as opposed to classes of leaf languages.

Acknowledgment. For helpful discussions I am grateful to J. Makowsky (Haifa) and H. Schmitz (Würzburg).

References

All96.   E. Allender. A note on uniform circuit lower bounds for the counting hierarchy. In *Proceedings 2nd Computing and Combinatorics Conference*, volume 1090 of *Lecture Notes in Computer Science*, pages 127–135. Springer Verlag, 1996.

Bar89.   D. A. Mix Barrington. Bounded-width polynomial size branching programs recognize exactly those languages in NC^1. *Journal of Computer and System Sciences*, 38:150–164, 1989.

Bar92.   D. A. Mix Barrington. Quasipolynomial size circuit classes. In *Proceedings 7th Structure in Complexity Theory*, pages 86–93. IEEE Computer Society Press, 1992.

BC94.   D. P. Bovet and P. Crescenzi. *Introduction to the Theory of Complexity*. International Series in Computer Science. Prentice Hall, London, 1994.

BCS92.   D. P. Bovet, P. Crescenzi, and R. Silvestri. A uniform approach to define complexity classes. *Theoretical Computer Science*, 104:263–283, 1992.

BDG95.   J. L. Balcázar, J. Díaz, and J. Gabarró. *Structural Complexity I*. Texts in Theoretical Computer Science. Springer Verlag, Berlin Heidelberg, 2nd edition, 1995.

BI94.   D. A. Mix Barrington and N. Immerman. Time, hardware, and uniformity. In *Proceedings 9th Structure in Complexity Theory*, pages 176–185. IEEE Computer Society Press, 1994.

BIS90.   D. A. Mix Barrington, N. Immerman, and H. Straubing. On uniformity within NC^1. *Journal of Computer and System Sciences*, 41:274–306, 1990.

BKS96.   B. Borchert, D. Kuske, and F. Stephan. On existentially first-order definable languages and their relation to NP. Technical Report MATH-AL-11-1996, Institut für Algebra, Technische Universität Dresden, 1996.

BL96.   B. Borchert and A. Lozano. Succinct circuit representations and leaf language classes are basically the same concept. *Information Processing Letters*, 58:211–215, 1996.

BLT92.   J. L. Balcázar, A. Lozano, and J. Torán. The complexity of algorithmic problems on succinct instances. In R. Baeza-Yates and U. Manber, editors, *Computer Science*. Plenum Press, New York, 1992.

Bor94.   B. Borchert. On the acceptance power of regular languages. In *Proceedings 11th Symposium on Theoretical Aspects of Computer Science*, volume 775 of *Lecture Notes in Computer Science*, pages 449–460. Springer Verlag, 1994.

BRS91.   R. Beigel, N. Reingold, and D. Spielman. PP is closed under intersection. In *Proceedings 23rd Symposium on Theory of Computing*, pages 1–9. ACM Press, 1991.

BS97.   B. Borchert and R. Silvestri. The general notion of a dot operator. In *Proceedings 12th Conference on Computational Complexity*, pages 26–44. IEEE Computer Society Press, 1997.

BT88.   D. A. Mix Barrington and D. Thérien. Finite monoids and the fine structure of NC^1. *Journal of the Association of Computing Machinery*, 35:941–952, 1988.

Bur96.   H. J. Burtschick. *Berechnungs- und Beschreibungskomplexität von Zählfunktionen und Lindströmquantoren*. PhD thesis, Fachbereich Informatik, TU-Berlin, 1996.

BV96.   H.-J. Burtschick and H. Vollmer. Lindström quantifiers and leaf language definability. Technical Report 96-005, Electronic Colloquium on Computational Complexity, 1996. Submitted for publication.

BVW96.   R. V. Book, H. Vollmer, and K. W. Wagner. On type-2 probabilistic quantifiers. In *Proceedings 23rd International Colloquium on Automata, Languages and Programming*, volume 1099 of *Lecture Notes in Computer Science*, pages 369–380. Springer Verlag, 1996.

BW96.   H. Baier and K. W. Wagner. The analytic polynomial-time hierarchy. Technical Report 148, Institut für Informatik, Universität Würzburg, 1996.

CC95.   L. Cai and J. Chen. On input read-modes of alternating turing machines. *Theoretical Computer Science*, 148:33–55, 1995.

CF91.   J.-Y. Cai and M. Furst. PSPACE survives constand-width bottlenecks. *International Journal of Foundations of Computer Science*, 2:67–76, 1991.

CHVW97.   K. Cronauer, U. Hertrampf, H. Vollmer, and K. W. Wagner. The chain method to separate counting classes. *Theory of Computing Systems*, 1997. To appear.

CMTV96.   H. Caussinus, P. McKenzie, D. Thérien, and H. Vollmer. Nondeterministic NC^1 computation. In *Proceedings 11th Computational Complexity*, pages 12–21. IEEE Computer Society Press, 1996.

Ebb85.   H.-D. Ebbinghaus. Extended logics: The general framework. In J. Barwise and S. Feferman, editors, *Model-Theoretic Logics*, Perspectives in Mathematical Logic, chapter II, pages 25–76. Springer Verlag, 1985.

EF95.   H.-D. Ebbinghaus and J. Flum. *Finite Model Theory*. Springer Verlag, Berlin Heidelberg, 1995.

Eil76.   S. Eilenberg. *Automata, Languages, and Machines*, volume B. Academic Press, New York, 1976.

Fag74.   R. Fagin. Generalized first-order spectra and polynomial time recognizable sets. In R. Karp, editor, *Complexity of Computations*, pages 43–73, 1974.

For97.   L. Fortnow. Counting complexity. In A. Selmand and L. A. Hemaspaandra, editors, *Complexity Theory Retrospective II*, pages 81–107. Springer Verlag, New York, 1997.

FSS84.   M. Furst, J. B. Saxe, and M. Sipser. Parity, circuits, and the polynomial-time hierarchy. *Mathematical Systems Theory*, 17:13–27, 1984.

Got95.   G. Gottlob. Relativized logspace and generalized quantifiers over finite structures. Technical Report CD-TR-95/76, Institut for Information Systems, Vienna University of Technology, 1995. An extended abstract appeared in the proceedings of the 10th Symposium on Logic in Computer Science, 1995.

Her95a.   U. Hertrampf. Classes of bounded counting type and their inclusion relations. In *Proceedings 12th Symposium on Theoretical Aspects of Computer Science*, volume 900 of *Lecture Notes in Computer Science*, pages 60–70. Springer Verlag, 1995.

Her95b.   U. Hertrampf. Regular leaf-languages and (non-) regular tree shapes. Technical Report A-95-21, Institut für Mathematik und Informatik, Medizinische Universität zu Lübeck, 1995.

Her97.   U. Hertrampf. Acceptance by transformation monoids (with an application to local self reductions). In *Proceedings 12th Conference on Computational Complexity*, pages 213–224. IEEE Computer Society Press, 1997.

HLS⁺93.   U. Hertrampf, C. Lautemann, T. Schwentick, H. Vollmer, and K. W. Wagner. On the power of polynomial time bit-reductions. In *Proceedings 8th Structure in Complexity Theory*, pages 200–207, 1993.

HVW95.   U. Hertrampf, H. Vollmer, and K. W. Wagner. On the power of number-theoretic operations with respect to counting. In *Proceedings 10th Structure in Complexity Theory*, pages 299–314, 1995.

HVW96.   U. Hertrampf, H. Vollmer, and K. W. Wagner. On balanced vs. unbalanced computation trees. *Mathematical Systems Theory*, 29:411–421, 1996.

Imm89.   N. Immerman. Expressibility and parallel complexity. *SIAM Journal on Computing*, 18:625–638, 1989.

JMT94.   B. Jenner, P. McKenzie, and D. Thérien. Logspace and logtime leaf languages. In *9th Annual Conference Structure in Complexity Theory*, pages 242–254, 1994.

Joh90.   D. S. Johnson. A catalog of complexity classes. In J. van Leeuwen, editor, *Handbook of Theoretical Computer Science*, volume A, pages 67–161. Elsevier, 1990.

KSV97.   S. Kosub, H. Schmitz, and H. Vollmer. Uniformly defining complexity classes of functions. Technical report, Institut für Informatik, Universität Würzburg, 1997.

Lan86.   K. J. Lange. Two characterizations of the logarithmic alternation hierarchy. In *Proceedings 12th Symposium on Mathematical Foundations of Computer Science*, volume 233 of *Lecture Notes in Computer Science*, pages 518–526. Springer Verlag, 1986.

Lin66.   P. Lindström. First order predicate logic with generalized quantifiers. *Theoria*, 32:186–195, 1966.

MP93. J. A. Makowsky and Y. B. Pnueli. Oracles and quantifiers. In *Computer Science Logic*, volume 832 of *Lecture Notes in Computer Science*, pages 189–222. Springer Verlag, 1993.

MP94. J. A. Makowsky and Y. B. Pnueli. Logics capturing relativized complexity classes uniform. In D. Leivant, editor, *Logic and Computational Complexity*, volume 1995 of *Lecture Notes in Computer Science*. Springer Verlag, 1994.

Pap94. C. H. Papadimitriou. *Computational Complexity*. Addison-Wesley, Reading, MA, 1994.

RS97. R. Rozenberg and A. Salomaa, editors. *Handbook of Formal Languages*, volume I. Springer Verlag, 1997.

RV97. K. Regan and H. Vollmer. Gap-languages and log-time complexity classes. *Theoretical Computer Science*, 1997. To appear.

Sip83. M. Sipser. Borel sets and circuit complexity. In *Proceedings of the 15th Symposium on Theory of Computing*, pages 61–69. ACM Press, 1983.

SM73. L. J. Stockmeyer and A. R. Meyer. Word problems requiring exponential time. In *Proceedings 5th ACM Symposium on the Theory of Computing*, pages 1–9, 1973.

Ste91. I. A. Stewart. Comparing the expressibility of languages formed using NP-complete operators. *Journal of Logic and Computation*, 1:305–330, 1991.

Ste92. I. A. Stewart. Using the Hamilton path operator to capture NP. *Journal of Computer and System Sciences*, 45:127–151, 1992.

Str94. H. Straubing. *Finite Automata, Formal Logic, and Circuit Complexity*. Birkhäuser, Boston, 1994.

Thé81. D. Thérien. Classification of finite monoids: the language approach. *Theoretical Computer Science*, 14:195–208, 1981.

Tod91. S. Toda. PP is as hard as the polynomial time hierarchy. *SIAM Journal on Computing*, 20:865–877, 1991.

Tor91. J. Toran. Complexity classes defined by counting quantifiers. *Journal of the ACM*, 38:753–774, 1991.

Vää94. J. Väänänen. A short course on finite model theory. Graduate Texts in Mathematics vol. 1, University of Helsinki, 1995.

Vei96. H. Veith. Succinct representation, leaf languages, and projection reductions. In *Proceedings 10th Structure in Complexity Theory*, pages 118–126. IEEE Computer Society Press, 1996.

Ver93. N. K. Vereshchagin. Relativizable and non-relativizable theorems in the polynomial theory of algorithms. *Izvestija Rossijskoj Akademii Nauk*, 57:51–90, 1993. In Russian.

Vol96a. H. Vollmer. Relations among parallel and sequential computation models. In *Proceedings 2nd Asian Computing Science Conference*, volume 1179 of *Lecture Notes in Computer Science*, pages 23–32. Springer Verlag, 1996.

Vol96b. H. Vollmer. Succinct inputs, Lindström quantifiers, and a general complexity theoretic operator concept. Technical Report 158, Institut für Informatik, Universität Würzburg, 1996.

Vol97. H. Vollmer. Relating polynomial time to constant depth. Technical Report 179, Institut für Informatik, Universität Würzburg, 1997.

VW97. H. Vollmer and K. W. Wagner. On operators of higher types. In *Proceedings 12th Conference on Computational Complexity*, pages 174–184. IEEE Computer Society Press, 1997.

Wag86a. K. W. Wagner. The complexity of combinatorial problems with succinct input representation. *Acta Informatica*, 23:325–356, 1986.

Wag86b.   K. W. Wagner. Some observations on the connection between counting and recursion. *Theoretical Computer Science*, 47:131–147, 1986.
Wra77.   C. Wrathall. Complete sets and the polynomial-time hierarchy. *Theoretical Computer Science*, 3:23–33, 1977.
Yao85.   A. C. C. Yao. Separating the polynomial-time hierarchy by oracles. In *Proceedings 26th Foundations of Computer Science*, pages 1–10. IEEE Computer Society Press, 1985.

Ramsey Theory Is Needed for Solving Definability Problems of Generalized Quantifiers *

Kerkko Luosto

Department of Mathematics, P.O.Box 4 (Yliopistonkatu 5)
00014 University of Helsinki, Finland
kerkko.luosto@helsinki.fi
http://www.logic.math.helsinki.fi

1 Introduction

In recent years, generalized quantifiers (see [H3]) have received quite a lot of novel interest because of their applications to computer science and linguistics. Their definability theory has made considerable progress during the last decade, which will be the subject of the next section. The proofs of many of these results often use results of Ramsey theory, such as theorems of van der Waerden and Folkman, and yet, the answers to some of the definability problems seem obvious from the outset. This raises the natural question whether Ramsey theory is really needed in the proofs (cf. [vBW]) or whether easier ways of proof might be discovered. The purpose of this paper is to argue in favour of the former and to convince the reader of the cruciality of Ramsey theory for quantifier definability theory.

The task of showing the necessity of Ramsey theory for quantifier theory does not translate easily to a rigorous mathematical problem, so the formulation of the framework is itself a problem. Roughly, we want to find an undefinability result $\mathcal{P}$ concerning quantifiers and a combinatorial theorem $\mathcal{K}$ and then show that $\mathcal{P} \Rightarrow \mathcal{K}$. We may as well fix that $\mathcal{K}$ is van der Waerden's theorem, whose variants have occured frequently in the recent quantifier literature. However, since $\mathcal{P}$ and $\mathcal{K}$ are both theorems of ZFC, the statement $\mathcal{P} \Rightarrow \mathcal{K}$ holds trivially, which is not really what we meant. This kind of difficulties are usually overcome in reverse mathematics by the metamathematical change to weaker axiom systems. However, I have abandoned this approach in order to preserve the clarity of the exposition. The reason is that generalized quantifiers are complex objects and the formulation of the basics of their theory in a weak axiom system would require a lot of tedious technical work.

Instead, a more combinatorial way of arguing is used. At first, a simple proof of $\mathcal{P} \Rightarrow \mathcal{K}$ is presented (in Section 4), which should already convince a full-fledged pragmatic. In essence, this proof shows that certain function related to quantifiers grow faster than van der Waerden's function W, which will be elaborated in Section 5. What exactly is meant, becomes more apparent as we proceed.

---

* 1991 *Mathematics Subject Classification*: 03C80, 05D10.

J. Väänänen (Ed.): ESSLLI'97, LNCS 1754, pp. 124–139, 1999.

We have not yet paid any attention to the choice of the undefinability result $\mathcal{P}$. Clearly it is not reasonable to allow $\mathcal{P}$ to be any quantifier result whatsoever, since it may be possible to code extremely difficult mathematical problems as quantifier problems; we should rather try to find a result $\mathcal{P}$ relevant for the field. As a result, we are not ready for the choice of $\mathcal{P}$ until in Section 4. Before that, we shall have a look at the problematics of quantifier definability theory in Section 2. We shall find out that we may as well restrict our considerations to monadic quantifiers, whose theory is sketched in the Section 3. The point is that, in the case of monadic quantifiers, the definability questions can be reduced to colouring problems about relations. These, in turn, give rise to certain fast-growing functions which give an upper bound for the van der Waerden's function.

2   Quantifier Definability Theory

Although 40 years has passed since Andrzej Mostowski presented his notion of a generalized quantifier, systematic treatment of definability problems of quantifiers is a fresh research subject. Up to the mid-1980's, the emphasis of the research in generalized quantifier theory was on finding logics with good model-theoretic properties. Definability and undefinability results were merely by-products of this study, e.g., Keisler's proof [K] that $\mathcal{L}_{\omega\omega}(Q_1)$ does not have the Δ-interpolation property was based on the following hidden result:

Theorem 1. (Keisler) Q_1^E *is not definable in* $\mathcal{L}_{\omega\omega}(Q_1)$. $\qquad\square$

To explain the quantifiers involved, let us introduce some notation. For a class of cardinals S, let C_S be the quantifier with defining class

$$K_{C_S} = \{\, \mathfrak{M} \in \mathrm{Str}(\{U\}) \mid |U^{\mathfrak{M}}| \in S \,\}$$

where U is a unary relation symbol. In other words, for every $\{U\}$-structure $\mathfrak{M}$ we have

$$\mathfrak{M} \models C_S x\, U(x) \iff |U^{\mathfrak{M}}| \in S.$$

Quantifiers of form C_S are called *cardinality quantifiers*. Similarly, E_S is the quantifier whose vocabulary is $\{E\}$ with E binary, and defining class K_{E_S} is the class of all structures $\mathfrak{M} \in \mathrm{Str}(\{E\})$ such that $E^{\mathfrak{M}}$ is an equivalence relation with the number of equivalence classes κ belonging to S. Clearly C_S is definable in $\mathcal{L}_{\omega\omega}(E_S)$ by the sentence

$$E_S x y\ (U(x) \wedge x = y).$$

Now $Q_1 = C_S$ and $Q_1^E = E_S$ where S is the class of all uncountable cardinals. Hence, Keisler's result shows that E_S is not necessarily definable in $\mathcal{L}_{\omega\omega}(C_S)$. His counterexample to Δ-interpolation of $\mathcal{L}_{\omega\omega}(Q_1)$ was later generalized by Xavier Caicedo [Ca2]. This, in turn, inspired some quantifier definability theory in finite model theory related to quantifiers similar to E_S by Flum, Schielen and Väänänen [FSV].

As it has become clear, the simplest form of a definability problem is the following:

(∗)   Is Q definable in the logic $\mathcal{L}_{\omega\omega}(\mathcal{Q})$?

where Q is a generalized quantifier and $\mathcal{Q}$ is a set of such. Most of the papers on generalized quantifiers where this is the main motivation are less than fifteen years old. One of the earliest is the paper by Luis Jaime Corredor [Co] solving completely the definability problems amongst cardinality quantifiers. Denote by $\oplus$ the common extension of integer and cardinal addition such that $\kappa \oplus n = n \oplus \kappa = \kappa$ for every infinite cardinal κ and $n \in \mathbb{Z}$. For S a class of cardinals and $n \in \mathbb{Z}$, let $S \oplus n = \{\kappa \oplus n \mid \kappa \in S\}$.

Theorem 2. (Corredor) *Let S and T_i, $i \in I$, be classes of cardinals. Then the following are equivalent:*
1) C_S is definable in $\mathcal{L}_{\omega\omega}(\{C_{T_i} \mid i \in I\})$.
2) There is a (finite) Boolean combination T of classes $T_i \oplus n$, $i \in I$ and $n \in \mathbb{Z}$, such that $|S \Delta T| < \omega$, i.e., the symmetric difference of S and T is finite. □

Observing that $\exists = C_E$ where E is the class of non-zero cardinals, case 2 can actually be replaced by
2') S is a Boolean combination of classes $T_i \oplus n$ and $E \oplus n$ where $i \in I$ and $n \in \mathbb{Z}$.
In the hindsight we may say that the neat formulation of the preceding theorem is possible because cardinality quantifiers enhance the expressive power of the first order logic by expressing things about extremely simple structures: the defining class of a cardinality quantifier consists of structures whose vocabulary contains only one predicate that is unary. In the same vein, it is reasonable to ask if theorem 1 is only an instance of a more general result, i.e., since the vocabulary of Q_1^E is $\{E\}$ with E binary and that of Q_1 is U with U unary, could it be possible that Q_1^E is not definable by any set $\mathcal{Q}$ of cardinality quantifiers. As it happens, this is exactly the case, and this is not accidental. For Q a quantifier with vocabulary τ, let the *arity* of Q be

$$\mathrm{ar}(Q) = \sup\{n_R \mid R \in \tau\}$$

where n_R is the arity of the relation symbol $R \in \tau$. A quantifier Q is *monadic*, if $\mathrm{ar}(Q) = 1$, *binary*, if $\mathrm{ar}(Q) = 2$, and *ternary*, if $\mathrm{ar}(Q) = 3$. The collection of all quantifiers Q with arity at most n is denoted by $\mathbf{Q}_n$. Then we have the following result due to Hella [H1] and implicitly by Caicedo [Ca1].

Theorem 3. *(Caicedo, Hella) Q_1^E is not definable by monadic quantifiers, i.e., in the logic $\mathcal{L}_{\omega\omega}(\mathbf{Q}_1)$, nor even in $\mathcal{L}_{\infty\omega}(\mathbf{Q}_1)$.* □

Even better, after the preliminary results of the other people (e.g., [G], [V1]) Hella was able to establish the following hierarchy result, among other results.

Theorem 4. *(Hella) For every $n \in \omega$ and non-zero ordinal α, the Magidor-Malitz quantifier Q_α^{n+1} is not definable in the logic $\mathcal{L}_{\infty\omega}(\mathbf{Q}_n)$.* □

The definition of Q_α^n is omitted; an interested reader may consult the original paper [H1] or the survey [HL].

At that time in the late 1980's, the focus in the research of generalized quantifiers was rapidly shifting towards finite structures. Hella's methods, certain model-theoretic games, were easily adaptable in the new context, as shown in the paper [H2]. Kolaitis and Väänänen [KV] did some systematic study in the realm of monadic simple quantifiers. A quantifier Q is *simple*, if its vocabulary consists of a single relation symbol. The *Härtig quantifier*, or the equicardinality quantifier, I, is an example of a monadic quantifier that is not simple; it is the monadic quantifier binding two formulas with the defining class

$$K_I = \{\, \mathfrak{M} \in \mathrm{Str}(\{U,V\}) \mid |U^{\mathfrak{M}}| = |V^{\mathfrak{M}}| \,\}.$$

The *Rescher quantifier* R has the same vocabulary as I, but its defining class is

$$K_R = \{\, \mathfrak{M} \in \mathrm{Str}(\{U,V\}) \mid |U^{\mathfrak{M}}| \leq |V^{\mathfrak{M}}| \,\}.$$

I is easily definable in $\mathcal{L}_{\omega\omega}(R)$ by the sentence

$$Rxy(U(x), V(y)) \wedge Rxy(V(x), U(y)).$$

Kolaitis and Väänänen proved the following:

Theorem 5. *(Kolaitis and Väänänen) Let $\mathcal{Q}$ be a finite set of simple monadic quantifiers. Then:*
a) I is not definable in $\mathcal{L}_{\omega\omega}(\mathcal{Q})$.
b) R is not definable in $\mathcal{L}_{\omega\omega}(\mathcal{Q} \cup \{I\})$.
c) $E_{2\mathbb{N}}$ is not definable in $\mathcal{L}_{\omega\omega}(\mathcal{Q} \cup \{I\})$ where $2\mathbb{N}$ is the set of even natural numbers.
These statements hold even if restricted to finite structures. $\square$

A notable feature in the proof is that it rests on the following results of Ramsey theory:

van der Waerden's Theorem. [vW] *For every $k, t \in \mathbb{N}$, there is $w \in \mathbb{N}$ such that if the set $\{0, \dots, w-1\}$ is coloured with at most t colours, say, by the colouring $\chi \colon \{0, \dots, w-1\} \to F$ with $|F| \leq t$, then there is a monochromatic arithmetic progression of length k, i.e., there are $a, d \in \mathbb{N}$, $d \neq 0$, such that $a + (k-1)d < w$ and $\chi(a) = \chi(a + id)$, for every $i = 0, \dots, k-1$.* $\square$

Folkman's Theorem. *For every $k, t \in \mathbb{N}$, there is $n \in \mathbb{N}$ such that for every colouring $\chi \colon \{0, \dots, w-1\} \to F$ with at most t colours (i.e., $|F| \leq t$), there exist $K \subset \mathbb{N}$ of size $|K| = k$ such that $\sum_{i \in K} i < n$ and for all non-empty $I, J \subset K$, we have $\chi(\sum_{i \in I} i) = \chi(\sum_{j \in J} j)$.* $\square$

These theorems give rise to combinatorial functions, e.g., van der Waerden's function $W \colon \mathbb{N} \times \mathbb{N} \to \mathbb{N}$ where for every $k, t \in \mathbb{N}$, $W(k, t)$ is the least $w \in \mathbb{N}$ satisfying the condition in the theorem. We also denote $W_2 \colon \mathbb{N} \to \mathbb{N}$, $W_2(n) = W(n, 2)$.

The reason why finite model theory pushed combinatorics in the front-line lies in the nature of the model constructions. While infinite structures can be made closed under different kinds of conditions and combinatorics remains hidden in the structures, this is usually not possible when constructing finite structures. This means that explicit use of combinatorial principles is required. In the sequel, Ramsey theory became a common tool in quantifier definability theory. The hierarchy result 4 inspired search for refinements of the hierarchy. In particular, linguistic interest (see [W]) seems to have spurred research on monadic quantifiers. The *width* of the quantifier Q is just the cardinality of its vocabulary τ, in symbols, $\mathrm{wd}(Q) = |\tau|$. Various researchers ([Li],[NV],[Lu]) obtained independently the following monadic hierarchy result:

Theorem 6. (Lindström, Nešetřil and Väänänen, Luosto) *For every $n \in \mathbb{N}$, there is a monadic quantifier of width $n + 1$ which is not definable by means of monadic quantifiers of width at most n, i.e., in any $\mathcal{L}_{\omega\omega}(\mathcal{Q})$ where $\mathcal{Q}$ is a set of monadic quantifiers of width at most n.* □

The explained involvement in combinatorics caught the eye of Johan van Benthem and Dag Westerståhl. They end the section 3 of their paper [vBW] by a short speculation of the need of the Ramsey theory. After a short description of the results and the methods in [KV], they write:

"This leads to the question whether every proof of the definability result requires Ramsey theory?"

And later:

"Given the very general nature of generalized quantifiers it may be worthwhile to do some 'reverse mathematics' in the field of finite combinatorics and definability questions, and thus to assess the *combinatorial content* of certain results about generalized quantifiers."

In order to be able to answer to these demands, we need to sketch the basics of monadic quantifier definability theory from [Lu] in the next section.

3   Reduction to Combinatorics

In this section some basics of the monadic quantifier definability theory are described, which is a prerequisite for the analysis of the combinatorial contents of undefinability results. The technicalities are kept aside, so that the presentation is necessarily sketchy and appeals a lot on the intuition of the reader. (The results and exact definitions can be found in [Lu].) Something concrete is needed for the further treatment, though, and this core of the theory is condensed in the notion of the irreducibility of a relation in the end of the section.

To start with, we note that monadic structures are easily describable. A structure with three unary predicates, say $\mathfrak{M} \in \mathrm{Str}(\{P, Q, R\})$, is drawn in the Figure 1 below.

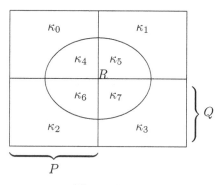

Figure 1

The structure $\mathfrak{M}$ is uniquely determined if we count the number of elements in each part, i.e., by the tuple $(\kappa_0, \ldots, \kappa_7)$. In general, a monadic structure $\mathfrak{M}$ for a finite vocabulary τ, say $|\tau| = l$, can be identified by a tuple $\kappa_{\mathfrak{M}}$ of length 2^l of cardinal invariants. In other words, if $\mathfrak{M}, \mathfrak{N} \in \mathrm{Str}(\tau)$, then $\mathfrak{M}$ and $\mathfrak{N}$ are isomorphic iff $\bar{\kappa}_{\mathfrak{M}} = \bar{\kappa}_{\mathfrak{N}}$.

Recall that for each cardinality quantifier Q there is a class S of cardinals such that $Q = C_S$. Analogously, each monadic quantifier Q with finite vocabulary can be reduced to a relation

$$\mathcal{R}(Q) = \{\, \bar{\kappa}_{\mathfrak{M}} \mid \mathfrak{M} \in K_Q \,\}.$$

Example 1. The vocabulary of the Härtig quantifier is $\{U, V\}$ and for $\mathfrak{M} \in \mathrm{Str}(\{U, V\})$, we have

$$\mathfrak{M} \in K_I \iff |U^{\mathfrak{M}}| = |V^{\mathfrak{M}}|.$$

To describe $\mathfrak{M} \in \mathrm{Str}(\{U, V\})$ up to isomorphism, we count that there are $\kappa_{0,\mathfrak{M}}$ elements in the intersection of the predicates, $U^{\mathfrak{M}} \cap V^{\mathfrak{M}}$, $\kappa_{1,\mathfrak{M}}$ elements in $U^{\mathfrak{M}} \smallsetminus V^{\mathfrak{M}}$, $\kappa_{2,\mathfrak{M}}$ elements in $V^{\mathfrak{M}} \smallsetminus U^{\mathfrak{M}}$ and $\kappa_{3,\mathfrak{M}}$ elements outside the predicates $U^{\mathfrak{M}}$ and $V^{\mathfrak{M}}$ (note that we have to fix some order in which to enumerate the invariants, but it is immaterial which particular order, as soon as it is the same for all structures). Therefore,

$$\bar{\kappa}_{\mathfrak{M}} = (|U^{\mathfrak{M}} \cap V^{\mathfrak{M}}|, |U^{\mathfrak{M}} \smallsetminus V^{\mathfrak{M}}|, |V^{\mathfrak{M}} \smallsetminus U^{\mathfrak{M}}|, |\mathrm{Dom}(\mathfrak{M}) \smallsetminus U^{\mathfrak{M}} \cup V^{\mathfrak{M}}|)$$

and

$$\mathfrak{M} \in K_I \iff \kappa_{0,\mathfrak{M}} \oplus \kappa_{1,\mathfrak{M}} = \kappa_{0,\mathfrak{M}} \oplus \kappa_{2,\mathfrak{M}}.$$

Hence,

$$\mathcal{R}(I) = \{\, (\kappa_0, \kappa_1, \kappa_2, \kappa_3) \mid \kappa_0 \oplus \kappa_1 = \kappa_0 \oplus \kappa_2 \,\}.$$

Note that for finite $\mathfrak{M} \in \mathrm{Str}(\{U, V\})$ we have the simple relation

$$\mathfrak{M} \in K_I \iff \kappa_{1,\mathfrak{M}} = \kappa_{2,\mathfrak{M}}.$$

Similarly, for the Rescher quantifier we get

$$\mathcal{R}(R) = \{ (\kappa_0, \kappa_1, \kappa_2, \kappa_3) \mid \kappa_0 \oplus \kappa_1 \leq \kappa_0 \oplus \kappa_2 \}$$

and

$$\mathcal{R}(R) \cap \mathbb{N}^4 = \{ (m_0, m_1, m_2, m_3) \in \mathbb{N}^4 \mid m_1 \leq m_2 \}.$$

There is a result analogous to Theorem 2 for cardinality quantifiers that characterizes the definability among monadic quantifiers. In order to explain the notation, let us consider the structure $\mathfrak{A}$ from Figure 1 and the sentence

$$\varphi = Ixy(P(x) \vee Q(x), R(y)).$$

Then $\mathfrak{A} \models \varphi$ iff $\left| P^{\mathfrak{A}} \cup Q^{\mathfrak{A}} \right| = \left| R^{\mathfrak{A}} \right|$ iff the interpreted structure $\mathfrak{M}$ is in K_I for which $U^{\mathfrak{M}} = P^{\mathfrak{A}} \cup Q^{\mathfrak{A}}$ and $V^{\mathfrak{M}} = R^{\mathfrak{A}}$.

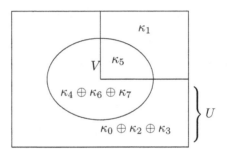

Figure 2

It is easily seen that $\bar{\kappa}_{\mathfrak{M}} = (\kappa_4 \oplus \kappa_6 \oplus \kappa_7, \kappa_0 \oplus \kappa_2 \oplus \kappa_3, \kappa_5, \kappa_1)$. Schematically we get $\bar{\kappa}_{\mathfrak{M}}$ from the components of $\bar{\kappa}_{\mathfrak{A}}$ when we sum over the set of indices $\{4, 6, 7\}$, $\{0, 2, 3\}$, $\{5\}$ and $\{1\}$, each in turn. In symbols, we write

$$\bar{\kappa}_{\mathfrak{M}} = \bar{s}(\bar{\kappa}_{\mathfrak{A}}, \overline{U})$$

where $\overline{U} = (\{4, 6, 7\}, \{0, 2, 3\}, \{5\}, \{1\})$. Then we have $\mathfrak{A} \models \varphi$ iff $\bar{s}(\bar{\kappa}, \overline{U}) \in \mathcal{R}(I)$.

Similarly to Theorem 2 we have to take translates into account. Below $J_{n,l}$ stands roughly for the set of all meaningful pairs $(\overline{U}, \bar{n})$ when we reduce the size of the tuple from n to l. Reformulating a result of Väänänen [V2] in our notation, we get:

Theorem 7. (Väänänen) *Let Q be a monadic quantifier and $\mathcal{Q}$ a set of monadic quantifiers, all with finite vocabularies. Let $n = 2^{\mathrm{wd}(Q)}$ and $l_q = 2^{\mathrm{wd}(q)}$, for each $q \in \mathcal{Q} \cup \{\exists\}$. Then the following are equivalent:*
1) Q is definable in $\mathcal{L}_{\omega\omega}(\mathcal{Q})$.

2) $\mathcal{R}(Q)$ is a Boolean combination of relations of form

$$\{\,\bar{\kappa} \mid \bar{s}(\bar{\kappa},\overline{U}) \oplus \bar{n} \in \mathcal{R}(q)\,\}$$

where $q \in \mathcal{Q} \cup \{\exists\}$ and $(\overline{U},\bar{n}) \in J_{n,l_q}$.   □

It is noteworthy that if we replace $\mathcal{R}(Q)$ by R and $\mathcal{R}(Q)$'s by some S_i's in case 2, what is left is a totally combinatorial condition on relations. This kind of consideration led to the notions of the rank $r(R)$ of a relation R and the relative rank $r_+(R)$ of a relation relative to a monoid $\langle M, +\rangle$ in [Lu]. The former is simpler, but the latter corresponds better to the intended application.

A quantifier Q with vocabulary τ is called *universe-independent*, if we have $\mathfrak{A} \in K_Q$ iff $\mathfrak{B} \in K_Q$ whenever $\mathfrak{A}, \mathfrak{B} \in \mathrm{Str}(\tau)$ are such that for every $R \in \tau$, it holds that $R^{\mathfrak{A}} = R^{\mathfrak{B}}$. Then the following hold:

Theorem 8. *Let $l \in \mathbb{N}$ and κ be an infinite cardinal and C be the set of cardinals below κ. Let Q be a monadic quantifier such that $n = 2^{\mathrm{wd}(Q)} \in \mathbb{N}$. Then the following are equivalent:*
1) Q is definable by monadic universe-independent quantifiers of width at most l on structures of cardinality less than κ.
2) $r_\oplus(\mathcal{R}(Q) \cap C^n) \leq 2^l - 1$.   □

Proposition 1. *Let $R \subset C^n$ be a relation where C is a set of cardinals. Then $r_\oplus(R) \leq r(R)$. If C is infinite, $C \cap \mathbb{N} = \{0\}$ and $r(R) > 3$, then $r_\oplus(R) = r(R)$.*
□

Let me sketch one application of these combinatorial results. Every quantifier has liftings which are called resumptions. For example, the second resumption of the Härtig quantifier is the quantifier $I^{(2)}$ with the vocabulary $\{R, S\}$ where R and S are binary and such that for every $\mathfrak{M} \in \mathrm{Str}(\{R, S\})$,

$$\mathfrak{M} \models I^{(2)}xy, tu(R(x,y), S(t,u))$$

iff

$$\left|R^{\mathfrak{M}}\right| = \left|S^{\mathfrak{M}}\right|,$$

i.e., if there are as many *pairs* in $R^{\mathfrak{M}}$ as there are pairs in $S^{\mathfrak{M}}$. Note that even if the definitions of I and $I^{(2)}$ have the same appearance, $I^{(2)}$ is a binary quantifier rather than monadic, since it binds two variables in both formulas.

Westerståhl posed, among other problems, the question if $I^{(2)}$ is definable in terms of monadic quantifiers of finite width. The question was answered to the negative in [Lu]. Indeed, suppose to the contrary that there is a finite set $\mathcal{Q}$ of monadic quantifiers of finite width such that $I^{(2)}$ is definable in $\mathcal{L}_{\omega\omega}(\mathcal{Q})$. Without loss of generality, these quantifiers are universe-independent. Pick $m \in \mathbb{N}$ such that $m \geq \mathrm{wd}(Q)$, for every $Q \in \mathcal{Q}$. Let $n = 2^m$. For $\bar{q} \in \mathbb{Z}^n$, consider the relation

$$R_{\bar{q}} = \{\,\bar{x} \in \mathbb{N}^n \mid \bar{x} \neq \bar{0}, \ \bar{q} \text{ and } \bar{x} \text{ orthogonal}\,\}.$$

(Here, orthogonality is just the usual notion of linear algebra, i.e., $\bar{q}$ and $\bar{x}$ are orthogonal iff $\bar{q} \cdot \bar{x} = 0$.) Then it can be shown that if $Q_{\bar{q}}$ is the quantifier with $\mathcal{R}(Q_{\bar{q}}) = R_{\bar{q}}$, then Q is definable in $\mathcal{L}_{\omega\omega}(I^{(2)})$ and therefore also in $\mathcal{L}_{\omega\omega}(\mathcal{Q})$. To demonstrate this by an example, consider $\bar{q} = (0, 3, -2, 0)$. Supposing that the vocabulary of the quantifier is $\{U, V\}$ and the indices as in Example 1, put

$$\exists y_0, y_1, y_2 \quad \exists t_0, t_1 \quad \Big(\neg y_0 = y_1 \wedge \neg y_0 = y_2 \wedge \neg y_1 = y_2 \wedge \neg t_0 = t_1 \wedge$$
$$I^{(2)} xy, zt \ \Big(U(x) \wedge \neg V(x) \wedge (x = x_0 \vee x = x_1 \vee x = x_2),$$
$$V(x) \wedge \neg U(x) \wedge (t = t_0 \vee t = t_1) \Big) \Big).$$

Then for every $\mathfrak{M} \in \mathrm{Str}(\{U, V\})$, $\mathfrak{M} \models \varphi$ iff $\mathfrak{M} \in K_{Q_{\bar{q}}}$ provided that there are at least three elements in $\mathfrak{M}$.

Now by the previous theorem, we have $r_{\oplus}(R_{\bar{q}}) \leq 2^m - 1 < n$. On the other hand, it may be shown that there are vectors $\bar{q}$ such that $r_{\oplus}(R_{\bar{q}}) = n$, which is a contradiction.

The notion of irreducibility is derived from the concept of the rank, not of that of relative rank. The preceding results show that the rank works well if we consider infinite structures. Since the emphasis of today's research is on finite structures, though, we might have a problem. The explanation why irreducibility is based on rank rather than the relative rank is postponed later.

Definition 1. *Let k and n be positive integers. The relation $R \subset A^n$ is k-reducible, if there are colourings $\xi_i \colon A^{n-1} \to F_i$ of A^{n-1} with at most k colours $(|F_i| \leq k)$, for $i = 0, \ldots, n-1$, such that the following holds: Let $\xi \colon A^n \to F_0 \times \cdots F_{n-1}, \xi(a_0, \ldots, a_{n-1}) = (c_0, \ldots, c_{n-1})$ where*

$$c_i = \xi_i(a_0, \ldots, a_{i-1}, a_{i+1}, \ldots, a_{n-1}).$$

(We call ξ the merge of ξ_i, $i = 0, \ldots, n-1$. We use this notion even if the sequence ξ_i is not complete.) Then for $\bar{a}, \bar{b} \in A^n$, if $\xi(\bar{a}) = \xi(\bar{b})$ and $\bar{a} \in R$, then also $\bar{b} \in R$. R is reducible if it is k-reducible for some $k \in \mathbb{N}^*$, otherwise R is irreducible.

Since understanding the definition of irreducibility is absolutely crucial for the purpose of this paper, let me repeat the definition in a different form: A relation R is reducible if there is a finite alphabet Σ and a language L of words of length n such that, given $\bar{a} \in A^n$, you can decide if $\bar{a} \in R$ by the following kind of a procedure. You first write down the tuple

$$\bar{a} = (a_0, \ldots, a_i, \ldots, a_{n-1})$$

on the paper. Then, for each $i = 0, \ldots, n-1$ consecutively, you first hide the component a_i, then take a look at the remaining tuple

$$\bar{a}' = (a_0, \ldots, a_{i-1}, a_{i+1}, \ldots, a_{n-1}),$$

write down your notes about the tuple, or actually just one letter $c_i \in \Sigma$, and then forget about the whole thing. After the process, you have written a word

$$w = c_0 c_1 \cdots c_{n-1}.$$

If $w \in L$, then $\bar{a} \in R$, otherwise $\bar{a} \notin R$.

Example 2. Let X be the set of all people and R the relation of all triples $(a, b, c) \in X^3$ such that a, b and c are sisters. Then R is reducible, even 2-reducible (actually all finite relations are reducible, so k-reducible for some k, but the parameter k there may vary). Indeed, first hide c and take a look at the pair (a, b). If a and b are sisters, write down Y(es), otherwise N(o). Proceed similarly, and if in the end you have the word YYY on the sheat, a and b and c are sisters.

The notion of irreducibility of n-ary R has been tailored equivalent to the statement $r(R) = n$. Let us then present a special case of Theorem 8. A monadic quantifier of finite width Q is called a *set quantifier* if $\mathcal{R}(Q)$ is a set, or equivalently $\mathcal{R}(Q) \subset C^n$ for some set C of cardinals and $n \in \mathbb{N}^*$. A monadic set quantifier Q of finite width is *biassed towards infinite*, if this C contains only infinite cardinals and possibly 0. Another way to put this is to say that there is an infinite κ such that for every $\mathfrak{M} \in K_Q$, the cardinality of $\mathfrak{M}$ is less than κ and every element a of $\mathfrak{M}$ has an infinite orbit under the automorhism group of $\mathfrak{M}$. Theorem 8, with the aid of Proposition 1 and the fact that irreducible relations are just those whose rank is the same as arity, implies the following theorem.

Theorem 9. *Let Q be a monadic set quantifier biassed to infinite with* $\mathrm{wd}(Q) \geq 2$. *Then the following are equivalent:*
1) Q is not definable in terms of monadic universe-independent quantifiers of width at most $\mathrm{wd}(Q)$.
2) $\mathcal{R}(Q)$ is irreducible. □

In the sequel, we need some technical results on irreducibility.

Lemma 1. *A non-unary relation $R \subset \mathbb{N}^n$ is irreducible iff for every $k \in \mathbb{N}^*$ there is $l \in \mathbb{N}$ such that $R \subset \{0, \ldots, l-1\} \subset \mathbb{N}^n$ is k-irreducible.*

Proof. Let us prove the equivalence of negations. Suppose first R is reducible. By definition, R is then k-reducible for some $k \in \mathbb{N}^*$ and for $i = 0, \ldots, n-1$, there are colourings $\xi_i \colon \mathbb{N}^{n-1} \to F_i$ such that for every $\bar{a}, \bar{b} \in \mathbb{N}^n$, if $\bar{a} \in R$ and $\xi(a) = \xi(b)$, then $\bar{b} \in R$, where ξ is the merge of the colourings ξ_i, $i = 0, \ldots, n-1$. Let $l \in \mathbb{N}$ be arbitrary. Pick a new colour c^* and put

$$\xi_i' \colon \mathbb{N}^{n-1} \to F_i \cup \{c^*\}, \xi_i'(\bar{a}) = \begin{cases} \xi_i(\bar{a}), & \text{for } \bar{a} \in \{0, \ldots, l-1\}^n \\ c^*, & \text{otherwise.} \end{cases}$$

Then clearly colourings ξ_i', $i = 0, \ldots, n-1$ show that $R \subset \{0, \ldots, l-1\} \subset \mathbb{N}^n$ is k+1-reducible.

To the other direction, suppose that there is $k \in \mathbb{N}^*$ such that for every $l \in \mathbb{N}$, the relation $R \subset \{0, \dots, l-1\}^n$ is k-reducible. We apply techniques from [LT]. Endowing the space $\mathcal{P}(\mathbb{N}^n)$ with its natural topology it can be shown that the set K ($K = K_n^n - 1, k, \mathbb{N}$ in the notation of [LT]) of k-reducible relations of $\mathbb{N}$ is a closed subset of $\mathcal{P}(\mathbb{N}^n)$. Now either for some $l \in \mathbb{N}$ it holds that $R = R \cap \{0, \dots, l-1\}^n$ or R is an accumulation point of $\{R \cap \{0, \dots, l-1\}^n \mid l \in \mathbb{N}\} \subset K$. In both of the cases, $R \in K$, i.e., R is k-irreducible. $\square$

Definition 2. *Let $R \subset \mathbb{N}^n$ be a non-unary irreducible relation. Then the complexity of irreducibility of R is $i_R \colon \mathbb{N}^* \to \mathbb{N}$,*

$$i_R(k) = \min\{\, l \in \mathbb{N} \mid R \cap \{0, \dots, l-1\}^n \subset \mathbb{N}^n \text{ is k-irreducible.}\,\}$$

4   Reverse Combinatorics: Proving van der Waerden's Theorem

We are finally in the position to choose the undefinability result $\mathcal{P}$ on which to base our analysis. For any ordinal α, denote $\mathrm{ind}(\aleph_\alpha) = \alpha$. For each $l \in \mathbb{N}^*$, let $n = 2^l$ and let S_l be the monadic quantifier with the vocabulary $\tau_n = \{U_0, \dots, U_{n-1}\}$ and with the defining class K_{S_l} consisting of $\mathfrak{M} \in \mathrm{Str}(\tau_n)$ such that all $U_i^{\mathfrak{M}}$, $i = 0, \dots, n-1$ are disjoint and infinite, but $\left|U_i^{\mathfrak{M}}\right| < \aleph_\omega$, and

$$\mathrm{ind}(|U_{n-1}^{\mathfrak{M}}|) = \sum_{i=0}^{n-2} \mathrm{ind}(|U_i^{\mathfrak{M}}|).$$

Let $\mathcal{P}_l$ be the following undefinability result (cf. [Lu, Example 4.6]).

$\mathcal{P}_l$: The quantifier S_l is not definable by means
of any universe-independent quantifiers of width l.

Moreover, let $\mathcal{P}$ be the statement: For every $l \in \mathbb{N}$, $\mathcal{P}_l$ holds.

Let us analyze quantifiers S_l in the spirit of the previous section. Let $\mathfrak{M} \in K_{S_l}$. If we enumerate the invariants in a suitable order, then

$$\bar{\kappa}_{\mathfrak{M}} = (\aleph_{m_0}, \dots, \aleph_{m_{n-1}}, 0, \dots, 0, \lambda)$$

with

$$m_{n-1} = \sum_{i=0}^{n-2} m_i$$

where for $i = 0, \dots, n-1$, $\aleph_{m_i} < \aleph_\omega$ is the number of elements in $U_i^{\mathfrak{M}}$, but outside other predicates, zeros refer to the empty intersections, and λ is the number of elements outside the predicates. Conversely, any $\mathfrak{M} \in \mathrm{Str}(\tau_l)$ with an invariant of this form is in K_{S_n}, so that

$$\mathcal{R}(S_n) = \{(\aleph_{m_0}, \dots, \aleph_{m_{n-1}}, 0, \dots, 0, \lambda) \mid m_0, \dots, m_{n-1} \in \mathbb{N},$$
$$\lambda \neq 0 \text{ or } m_{n-1} \neq 0, \text{ and } m_{n-1} = \textstyle\sum_{i=0}^{n-2} m_i\}.$$

Constant components and vacuous variables (such as λ) do not have any effect on the ranks of relations (cf. [Lu]), nor does the replacement of $\aleph_j$ by j. So for every $l \in \mathbb{N}^*$ and $n = 2^l$, $\mathcal{P}_n$ is equivalent to the following statement:

$\mathcal{P}'_n$: The relation

$$T_n = \{ (m_0, \dots, m_{n-1}) \in \mathbb{N}^n \mid m_{n-1} = \sum_{i=0}^{n-2} m_i \}$$

is irreducible.

Where does the relevance of this result lie? Firstly, $\mathcal{P}$ gives an easy way to prove the monadic hierarchy theorem, i.e., the fact that extending the width has the capacity of enhancing the expressive power of monadic quantifiers. Since this motive comes from inside the field rather than from applications, let me point out that in the notation of the previous section, $T_n = R_{\bar{q}}$ with $\bar{q} = (1, 1, , \dots, 1, -1) \in \mathbb{Z}^n$, for every $n \in \mathbb{N}$, $n \geq 2$, so that this problematics is related to the proof that $I^{(2)}$ is not definable by monadic quantifiers. Unfortunately, irreducibility does not mean that the simple relations T_n were useful for that particular proof. As it happens, $r_\oplus(T_n) = 2$ albeit $r(T_n) = n$, which illustrates well the difference between finite and infinite arithmetics. In spite of this, we may view it as a mere pragmatic simplification that we use the result $\mathcal{P}$ rather than undefinability of $I^{(2)}$ as the basis of our analysis. Indeed, we could have introduced the notion of irreducibility for the relative rank, and then we would have got functions $i_R^\oplus$ satisfying $i_R^\oplus(n) \geq i_R(n)$, for every $n \in \mathbb{N}^*$, and we could have gone through an analysis in the similar vein for relations of form $R_{\bar{q}}$. The trade-off would have been bad in the sense that we ought to have dealt with more complicated notions, with linear-algebraic technicalities (cf. [Lu, Section 5]) and in the end, the results would have been almost the same as we shall get now.

Abandoning this side-track, we first prove a technical lemma concerning partial functions. For tuples $\bar{a}$ and $\bar{b}$, let $\bar{a}\,\hat{}\,\bar{b}$ be the concatenation of $\bar{a}$ and $\bar{b}$.

Lemma 2. *Let* $k, n \in \mathbb{N}^*$ *and let* f *be an* n-*ary partial function on* A *which is* $k^n + 1$-*irreducible. Let* $\xi_i \colon A^n \to F_i$, $i = 0, \dots, n-1$, *be finite colourings with at most* k *colours. Let* ξ *be the merge of* ξ_i, $i = 0, \dots, n-1$. *Then there are* $\bar{a} = (a_0, \dots, a_{n-1}) \in \mathrm{dom}(f)$ *and* $c \in A$ *such that* $c \neq f(\bar{a})$ *and* $\xi(\bar{a}\,\hat{}\,(f(a))) = \xi(\bar{a}\,\hat{}\,(c))$, *i.e.,* $\xi(a_0, \dots, a_{n-1}, f(a)) = \xi(a_0, \dots, a_{n-1}, c)$.

Proof. Pick a new colour c^* and put $F_n = (F_0 \times \cdots \times F_{n-1}) \cup \{c^*\}$ and

$$\xi_n \colon A^n \to F_n, \; \xi_n(\bar{a}) = \begin{cases} \xi(\bar{a}\,\hat{}\,(f(\bar{a}))), & \text{for } \bar{a} \in \mathrm{dom}(f) \\ c^*, & \text{otherwise} \end{cases}$$

Then $|F_n| \leq k^n + 1$. Let ξ^* be the merge of ξ_i, $i = 0, \dots, n$. Since f is $k^n + 1$-irreducible, ξ^* cannot carry all the information to determine if a tuple is in f, so there are $\bar{a}, \bar{b} \in A^n$, $c, d \in A$ such that 1) $\xi^*(\bar{a}\,\hat{}\,(c)) = \xi^*(\bar{b}\,\hat{}\,(d))$ and 2) $d = f(\bar{b})$ and 3) either $\bar{a} \notin \mathrm{dom}(f)$ or $c \neq f(\bar{a})$. As $\bar{b} \in \mathrm{dom}(f)$, we have $\xi_n(\bar{b}) = \xi_n(\bar{a})$, whence the former possibility in case 3 is outruled. But $\xi_n(\bar{a}) = \xi_n(\bar{b})$ says even more, namely that $\xi(\bar{a}\,\hat{}\,(f(\bar{a}))) = \xi(\bar{b}\,\hat{}\,(f(\bar{b})))$. On the other hand, $\xi^*(\bar{a}\,\hat{}\,(c)) = \xi^*(\bar{b}\,\hat{}\,(d))$ implies $\xi(\bar{a}\,\hat{}\,(c)) = \xi(\bar{b}\,\hat{}\,(d))$. Hence, $\xi(\bar{a}\,\hat{}\,(c)) = \xi(\bar{a}\,\hat{}\,(f(a)))$.   □

Proposition 2. *Let $n \in \mathbb{N}$ with $n \geq 2$ and $\varrho\colon \mathbb{N}^n \to F$ be a finite colouring. Then the irreducibility of T_{n+1} implies that there are $\bar{a} \in \mathbb{N}^n$ and $d \in \mathbb{Z} \setminus \{0\}$ such that $\varrho(\bar{a}) = \varrho(\bar{a} + d\bar{e}_k)$, for $k = 0, \ldots, n-1$, where $\bar{e}_k$ is the unit vector having k^{th} component 1.*   □

Proof. Pick a new colour c^*. For each $i = 0, \ldots, n-1$, colour

$$\bar{a}'_i = (a_0, \ldots, a_{i-1}, a_{i+1}, \ldots, a_n)$$

by c^* if $b = a_n - \sum_{j=0,\, j\neq i}^{n-1} a_j < 0$, otherwise colour $\bar{a}'_i$ by

$$\varrho(a_0, \ldots, a_{i-1}, b, a_{i+1}, \ldots, a_{n-1}).$$

Let the resulting colouring be $\xi_i\colon \mathbb{N}^n \to F \cup \{c^*\}$. We can think of this colouring as an attempt to first recover the hole (i^{th} component) in the tuple and then to use colouring ϱ. Now, by the previous lemma, the merge ξ of ξ_i, $i = 0, \ldots, n-1$, does not carry all the information about T_n. Consequently, there are $\bar{a} = (a_0, \ldots, a_{n-1}) \in \mathbb{N}^n$ and $d \in \mathbb{Z}$, $d \neq 0$, such that $\xi(\bar{a}\,\hat{}\,(s)) = \xi(\bar{a}\,\hat{}\,(s+d))$ where $s = \sum_{i=0}^{n-1} a_i$. Unfolding the definition of the colourings, we see that for every $i = 0, \ldots, n-1$,

$$\begin{aligned} \varrho(\bar{a}) &= \xi_i(a_0, \ldots, a_{i-1}, a_{i+1}, \ldots, a_{n-1}, s) \\ &= \xi_i(a_0, \ldots, a_{i-1}, a_{i+1}, \ldots, a_{n-1}, s+d) = \varrho(\bar{a} + d\bar{e}_i). \end{aligned}$$ □

Now the goal is at hand:

Theorem 10. *(van der Waerden) Let $\chi\colon \mathbb{N} \to F$ be a finite colouring and $k \in \mathbb{N}$. Then there are $a, d \in \mathbb{N}$ such that $d \neq 0$ and $\chi(a) = \chi(a + id)$, for every $i = 0, \ldots, k-1$.*

Proof. We may assume that $k \geq 2$. Let $n = k-1 \in \mathbb{N}^*$ and consider the coloring

$$\varrho\colon \mathbb{N}^n \to F, \quad \varrho(x_0, \ldots, x_{n-1}) = \chi\left(\sum_{j=0}^{n-1}(j+1)x_j\right).$$

The previous proposition implies that there are $\bar{a} = (a_0, \ldots, a_{n-1}) \in \mathbb{N}^n$ and $d \in \mathbb{Z} \setminus \{0\}$ such that $\varrho(\bar{a}) = \varrho(\bar{a} + d\bar{e}_i)$, for every $i = 0, \ldots, n-1$. Consequently, for $a = \sum_{j=0}^{n-1}(j+1)a_j$ and $i = 0, \ldots, n-1$ we have

$$\chi(a) = \varrho(\bar{a}) = \varrho(\bar{a} + d\bar{e}_i) = \chi(a + (i+1)d).$$

Equivalently, for every $i = 0, \ldots, k-1$, it holds that $\chi(a) = \chi(a + id)$. If it happens that $d < 0$, the numbers $a' = a + (k-1)d$ and $d' = -d > 0$ fulfil the claim instead of a and d.   □

5   Fast-Growing Functions

As it was mentioned in the introduction, the ideas of the previous section can be converted to results on combinatorial functions. One of the most interesting of such is the van der Waerden's function $W \colon \mathbb{N} \times \mathbb{N} \to \mathbb{N}$. For some time there was speculation over whether W is even a primitive-recursive function until Saharon Shelah [S] proved that W is actually quite low in the Grzegorczyk (or Ackermann) hierarchy. Still, it is possible that W_2 grows faster than the exponential tower, i.e., for every $n \in \mathbb{N}$,

$$W_2(n) \geq \left. 2^{2^{\cdot^{\cdot^{\cdot^2}}}} \right\} n \text{ times}.$$

The last step of the paper is to link complexities of irreducibilities to the discussion by proving an upper bound for W. It is worth mentioning here that a fast-growing complexity of irreducibility i_R of R means intuitively that the irreducibility of R is difficult to discover, R is nearly reducible.

Theorem 11. *For every* $k, n \in \mathbb{N}$, $k, n \geq 2$, *we have that*

$$W(n, k) \leq (n-1) i_{T_n}((k+1)^{n-1} + 1).$$

Proof. Denote $t = i_{T_n}((k+1)^{n-1}+1)$ and $w = (n-1)t$. Let $\chi \colon \{0, \dots, w-1\} \to F$ be an arbitrary colouring with $|F| \leq k$. We need to show that $\{0, \dots, w-1\}$ includes a monochromatic progression of length k. For that purpose, we do the same tricks as in the previous section and consider the colouring $\varrho \colon \{0, \dots, t-1\} \to F \cup \{c^*\}$,

$$\varrho(a_0, \dots, a_{n-2}) = \begin{cases} \chi(\sum_{l=0}^{n-2}(l+1)a_l), & \text{if defined} \\ c^*, & \text{otherwise} \end{cases}$$

where c^* is a new colour. For each $i = 0, \dots, n-2$ and

$$\bar{a}' = (a_0, \dots, a_{i+1}, a_{i+1}, \dots, a_{n-1}) \in \mathbb{N}^n,$$

let $\bar{b}_i(\bar{a}') = (a_0, \dots, a_{i-1}, a_{n-1} - \sum_{l=0, l\neq i}^{n-2}(l+1)a_l, \dots, a_{n-2})$ be the tuple we can recover from $\bar{a}$, and let

$$\xi_i \colon \{0, \dots, t-1\}^{n-1} \to F \cup \{c^*\}, \ \xi_i(\bar{a}') = \begin{cases} \varrho(\bar{b}_i(\bar{a}')), & \text{if defined} \\ c^* & \text{otherwise} \end{cases}$$

Note that $T' = T_n \cap \{0, \dots, t-1\}^n$ is a partial function from $\{0, \dots, t-1\}^{n-1}$ to $\{0, \dots, t-1\}$. By the choice of t, T' is $(k+1)^{n-1}+1$-irreducible, so applying lemma 2 to T' and the merge ξ of ξ_i, $i = 0, \dots, n-2$, we find $\bar{a} = (a_0, \dots, a_{n-2}) \in \{0, \dots, t-1\}^{n-1}$ and $c \in \{0, \dots, t-1\}$ such that for $s = \sum_{i=0}^{n-2} a_i$, we have that

$c \neq s < t$ and $\xi(\bar{a}\,\hat{}\,(s)) = \xi(\bar{a}\,\hat{}\,(c))$. Observe that $u = \sum_{l=0}^{n-2}(l+1)a_l \leq (n-1)s < (n-1)t = w$. Hence, $\varrho(\bar{a}) = \chi(u)$ and for every $i = 0, \ldots, n-2$,

$$\xi_i(a_0, \ldots, a_{i-1}, a_{i+1}, \ldots, a_{n-2}, s)$$
$$= \varrho(a_0, \ldots, a_{i-1}, a_i, a_{i+1}, \ldots, a_{n-2})$$
$$= \varrho(\bar{a}) = \chi(u).$$

On the other hand, $\chi(\bar{a}\,\hat{}\,(s)) = \chi(\bar{a}\,\hat{}\,(c))$ implies that

$$\xi_i(a_0, \ldots, a_{i-1}, a_{i+1}, \ldots, a_{n-2}, c)$$
$$= \varrho(a_0, \ldots, a_{i-1}, a_i + d, a_{i+1}, \ldots, a_{n-2})$$
$$= \varrho(\bar{a} + d\bar{e}_i) = \varrho(\bar{a})$$

where $d = c - s \neq 0$. Hence $\chi(u) = \chi(u + id)$, for every $i = 0, \ldots, n-1$, so that $\{0, \ldots, w-1\}$ includes a monochromatic aritmetic progression of length n. Consequently, $W(n, k) \leq w$. $\qquad\square$

References

Ca1. Xavier Caicedo: *Maximality and interpolation in abstract logics (back-and-forth techniques)*. Doctoral Dissertation, University of Maryland, 1978, 146 pp.

Ca2. Xavier Caicedo: *Definability properties and the congruence closure.* **Archive for Mathematical logic** 30 (1990), 231–240.

Co. Luis Jaime Corredor: *El reticulo de las logicas de primer orden con cuantificadores cardinales.* **Revista Colombiana de Matemáticas** XX (1986), 1–26.

FSV. J. Flum, M. Schiehlen and J. Väänänen: *Quantifiers and congruence closure.* Studia Logica 62 (1999), 315–340.

G. S. C. Garavaglia: *Relative strength of Malitz quantifiers.* **Notre Dame Journal of Formal Logic** 19 (1978), 495–503.

H1. Lauri Hella: *Definability hierarchies of generalized quantifiers.* **Annals of Pure and Applied Logic** 43 (1989), 235–271.

H2. Lauri Hella: *Logical hierarchies in PTIME.* **Information and Computation** 129 (1996), 1–19.

H3. Lauri Hella: *Generalized quantifiers in finite model theory.* Manuscript, 17 pp.

HL. L. Hella and K. Luosto: *Finite generation problem and n-ary quantifiers.* In M. Krynicki, M. Mostowski and L. W. Szczerba (eds.): **Quantifiers: Logics, Models and Computation**, Kluwer Academic Publishers 1995, 63–104.

K. H. Jerome Keisler: *Logic with the quantifier "there exist uncountably many".* **Annals of Mathematical Logic** 1 (1970), 1–93.

KV. P. Kolaitis and J. Väänänen: *Generalized quantifiers and pebble games on finite structures.* **Annals of Pure and Applied Logic** 74 (1995), 23–75.

Li. Per Lindström: Personal communication. Via Dag Westerståhl, 1993.

Lu. Kerkko Luosto: *Hierarchies of monadic generalized quantifiers.* To appear in **The Journal of Symbolic Logic** .

LT. K. Luosto and Jerzy Tyszkiewicz: *On resumptions of monadic generalized quantifiers.* Manuscript, 6 pp.

NV. J. Nešetřil and Jouko Väänänen: *Combinatorics and quantifiers.* **Commentationes Mathematicae Universitatis Carolinae** 37 (1996), 433–443.

S. Saharon Shelah: *The theorems of Beth and Craig in abstract model theory, III: Δ-logics and infinitary logics.* **Israel Journal of Mathematics** 69, no.2 (1990), 193–213.

V1. Jouko Väänänen: *A hierarchy theorem for Lindström quantifiers.* In M. Furberg, T. Wetterström and C. Åberg (eds.): **Logic and Abstraction**. Acta Philosophica Gothoburgesia 1 (1986), 317–323.

V2. Jouko Väänänen: *Unary quantifiers on finite models.* **Journal of Logic, Language and Information** 6 (1997), 275-304.

vBW. J. van Benthem and D. Westerståhl: *Directions in generalized quantifier hierarchy.* **Studia Logica** 55 (1995), 389–419.

vW. B. L. van der Waerden: *Beweis einer Baudetschen Vermutung.* **Nieuw Arch. Wisk.** 15 (1927), 212–216.

W. Dag Westerståhl: *Quantifiers in natural language. A survey of some recent work.* In M. Krynicki, M. Mostowski and L. W. Szczerba (eds.): **Quantifiers: Logics, Models and Computation**, Kluwer Academic Publishers 1995, 359–408.